ISBN 978-0-9878109-5-3

Cover design by: OneFish Creative Inc.

*I dedicate this book to the tens of thousands
of Albertans, and other Canadians, who have
tapped the oil sands for the service of humanity.*

CONTENTS

INTRODUCTION

Few historians are able to participate in the events they describe and even fewer get to taste the blood, sweat and tears. I was fortunate to do both. For fourteen years I was communications head at Syncrude Canada and got to participate in that great mega-project's beginnings. This gave me a lasting appreciation for the personalities and forces that left their imprint on the oil sands industry and our nation.

In my view the creation of Syncrude, and the oil sands enterprises that followed, was a fine work and something of which Canadians should be proud. I appreciate that others may disagree, and some of them have written books and articles to say so. This is my take on it, and it is formed from experience and the best data I could find. I am not quite the "last man standing" from those early days of the industry (and I don't want to be)…but I'm near the front of the line.

This book is not a love letter to the oil industry, but it does unabashedly honour the good things the oil sands have provided the modern world. It *is* a love letter of sorts to the people of Canada who built and sustained the industry.

I left the industry more than 35 years ago, but I carried my interest in it through the rest of my career. In 2019, in the wake of the great upsurge of global controversy about the oil sands, I decided to pull together an account of the industry's birth and explosive growth and its dramatic collision with the climate change movement. Since every good story needs an ending, I decided to examine what the industry's possible futures may hold. I make no secret of my biases, but I have endeavored to be as fair as possible.

For the record I received no support, financial or otherwise, from the industry, the Alberta Government, or any U.S. or Canadian foundation, and requested none. I did receive wise advice and counsel from several retired industry executives but none from any current industry executive. Perhaps the reticence of current executives should not have surprised me since, more than five years ago, Chris Turner, the author of *The Patch: The People, Pipelines and Politics of the Oil Sands*, noted that "many of [the industry's] key players and some of its most vocal critics proved not just reluctant but completely unwilling to speak to me for this book."

Current industry leaders can rest easy. Whether my conclusions prove right or wrong, they are mine.

John J. Barr

FOREWORD

Oil Changes the World

At first, the 1858 discovery of so-called rock oil seeping from the ground near Oil Springs, Ontario, seemed at a minor event in the world. Hardly more than a curiosity, really. Apart from liniment, the only thing rock oil was thought to be good for was lamp fuel, and for that purpose, it was sold by the pail.

In 1870, John D. Rockefeller founded the Standard Oil Trust, and the first great modern oil company began to take shape. But at the beginning, Standard Oil didn't have a major customer to power its growth. That began to change 15 years later when a German inventor, Carl Benz, applied for a patent for his "vehicle powered by a gas engine," by which he meant the modern automobile.

Together, the oil and automobile industries powered the beginning of the Oil Age. Oil was to change the planet, replacing coal as the dominant form of energy for heating, industrial practices, and transportation by sea, land, and air. In short order, gasoline and other oil derivatives replaced coal as the dominant fuel for ships, cars, trucks, and airplanes.

Oil had two immensely useful properties: it was much more energy-dense than coal, charcoal, or wood (42 gigajoules per tonne compared with 25–27 megajoules per tonne for bituminous coal and 15–17 megajoules per tonne for wood). And it was immensely more energy-dense than wind or sun.[1]

Early in the 20th century, the centre of oil production began its shift from North America to the Middle East. In 1910, Winston Churchill, then a rising young politician and First Lord of the Admiralty, directed the Royal Navy to convert its operations from coal to oil. Four years later, the British government invested about $250 million to buy what was known as the D'Arcy Concession—an enormous oil-rich region now part of modern Iran. William Knox D'Arcy negotiated the concession from Mozzafar al-Din, shah of Persia. It gave him a monopoly on Persia's immense petroleum reserves and the exclusive rights to prospect for oil in Persia. The purchase eventually led to the creation the Anglo-Persian Oil Company, later British Petroleum or BP. By the First World War, the age of oil was well underway.

* * *

The purchase of the D'Arcy Concession was the first move of a Western government to align with Middle Eastern oil production and make oil a strategic commodity and principal driver of the modern economy. For Britain, its immediate consequences were hugely positive. From the 1920s into the 1940s, Britain's standard of living was supported by oil from Iran. British cars, trucks, and buses ran on cheap Iranian oil. Factories throughout Britain were fueled by oil from Iran. The Royal Navy, which projected British power all over the world, powered its ships with Iranian oil.

For the rest of the world, the geopolitical consequences were mixed. As one example, in 1953, determination to control Iran's oil led the American government to direct the CIA to overthrow nationalist Prime Minister Mohammed Mossadegh and install Mohammad Reza Shah Pahlavi. This would eventually trigger the Islamic Revolution, bringing the ayatollahs to power and influencing the tumultuous politics of the Middle East until this day.

More globally, it'd be difficult to exaggerate the impact of oil on humanity and life itself. Petroleum—this distillation of the remains of prehistoric animals, plants, and ancient solar energy—was an enormously useful, energy-dense liquid fuel that could easily be stored or transported and scaled to power everything from sea, land, and air travel to heating living rooms and going to Mars. It made possible global trade, international travel, mechanized industrial warfare, and, yes, ultimately the movement of millions of human beings around the world. The economist Vaclav Smil has written that the advent of motorized warfare changed the perception of petroleum from a useful commodity to a vital national resource for warfare.

Oil energy reduced an immense amount of human drudgery and enabled a vast expansion of human freedom to move and grow. For hundreds of millions of people, it ended what Karl Marx called "the idiocy of rural life." It created vast economic opportunities where there'd been none before, and it spawned millions of careers never before imagined.

But of course, it came with a price. It transformed landscapes, and while it shrank the world, it also caused a good deal of pollution as well. It propelled urban sprawl and helped generate smog and the entombment of vast stretches of fertile farmland under pavement. It made it possible for millions of people to "ride the broad highway in your Chevrolet"; it also made possible tank warfare and intercontinental ballistic missiles. It expanded people's freedom to discover the world, but it also expanded our ability to degrade the planet. It warmed cold places, but it also contributed to climate change, with consequences yet to be fully understood.

And it changed our culture, beginning with the livelihood and politics of a part of Canada called Alberta. This book is about some of those changes.

How Oil Changed Alberta and the Nation

For almost a century—from the discovery of oil in Ontario in 1858 to the settlement of Alberta and its development until the end of the Second World War—oil had little direct impact on Alberta's people and economy.

True, in 1914, drillers found a significant reservoir of highly pressurized natural gas and light oil in the foothills near Turner Valley, southwest of Calgary, but at the time, people scarcely knew what to do with it, particularly the natural gas. So much of the gas was flared—burnt off in flames that could be seen as far as 50 miles away—and its withdrawal depressurized the reservoir, leaving much of the oil in the ground.

For all intents and purposes, until the middle of the 20th century, Alberta was much like its sister provinces, Saskatchewan and Manitoba, a land of farmers and ranchers. People made their living in a boom-and-bust economy buffeted by fluctuating prices of grain and meat. Ordinary people worked hard for limited incomes and limited opportunities. The vast majority of Albertans lived constricted lives with very narrow personal horizons.

The province's politics was dominated by the fact that it had a small population and was part of Canada's so-called hinterland. Political and economic powers were rooted in central Canada, which elected federal governments, owned the factories and banks, and made both the decisions and the rules.

The Great Depression brought Alberta's semi-colonial status home to the people in a particularly painful way. Of course, the Depression was hard on *all* Canadians, but for the people of the Prairies, it was especially cruel. When the foundation of your economy is producing grain and meat and the price of both collapses, the oxygen is sucked out of every room.

My maternal grandmother was abandoned by her husband in 1931 and had to raise a family of five children as a part-time rural schoolteacher at a time when many rural school districts were broke and paid their teachers—when they could pay them anything—with chickens and bags of grain.. She and her children wore hand-me-downs and lived off their garden and a milk cow. My grandmother couldn't afford a car until 1952, and for the last 10 years of her life she lived in a one-room cabin behind my aunt's farmhouse.

Before the discovery of oil, people in Alberta and Saskatchewan tried to build a better world through politics. The provinces gave rise to two

Prairie protest movements: in Saskatchewan, a social-democracy movement called the Co-Operative Commonwealth Federation (CCF), and in Alberta, a monetary reform movement called Social Credit.

In its *Regina Manifesto*, the CCF proposed creating a "planned economy" based on the nationalization of banks, insurance companies, public utilities, mining, pulp and paper "and the distribution of milk, bread, coal and gasoline." A "national planning secretariat" of "competent managements freed from day to day political interference" would guide the economy to a better world.[2]

Social Credit, meanwhile, proposed to end poverty by nationalizing the banks and using them to provide every citizen $25 a month in purchasing power. Nationalizing the banks was the one policy that the two movements agreed on. Apart from that, they were rivals and fought tooth and nail for years. Social Credit governed for 27 years in Alberta, and the CCF was elected for 20 years in Saskatchewan. Neither was ever elected nationally.

All Social Credit's seize-the-banks initiatives were stymied by the courts or the federal government, which was effectively controlled by central Canada. After the war, Social Credit morphed into a middle-of-the-road "good government" party.

What brought the Depression to an end on

the Prairies wasn't politics but rearmament and the Second World War. Following the end of the war, however, there was little hope that the prosperity Alberta enjoyed in the early years of the century would return. And then, in 1947, after drilling a hundred dry holes, Imperial Oil found a major oil reservoir near the town of Leduc, south of Edmonton, and Alberta's world changed.

In less than 10 years, the province shifted from have-not to have. Social change accelerated. People moved to cities, and cities boomed. Revenues from oil poured into government and funded an enormous number of schools, universities, and hospitals. Teacher salaries soared. Thousands of young people flooded into post-secondary education. A modern intellectual and managerial class took root. For thousands of people, whole new careers opened up. Everything was made possible, in one way or another, by oil.

That didn't mean that ordinary people fell in love with oil, oilmen, or the physical manifestations of the industry. Nobody would ever confuse a tank farm, pumpjack, or oil refinery with public art. (A grain elevator is a different matter. My personal favourite was the old Alberta Wheat Pool elevator near Nisku on which someone had painted, in letters 10 feet high: *"REPENT! OR SPEND ETERNITY IN A LAKE OF EVERLASTING FIRE!"*)

If Albertans didn't love the sights, sounds, smells, or employees of the oil industry—rig workers called themselves "roughnecks" and often lived up to the name—they knew where their better lives came from. Oil paid for jobs, careers, schools, hospitals, and universities, all of which had been previously scarce. Oil gave Albertans opportunity. It underwrote the arts and built theatre centres.

It also gave people increased confidence and belief in their ability to make their own way in life. The writers of *Promised Land*, a Gus Van Sant film about the fracking industry in rural Pennsylvania, called oil and gas income "fuck-you money." In the film, Matt Damon plays a sales representative trying to get farmers to sell him drilling rights to their land. In a bar, he confronts farmers who just don't "get" what he's offering:

> I'm not talking about little pay increases. I'm talking about fuck-you money. You don't want to apply for college loans for your kids? This money says, "Fuck you, loans." You worried about car payments? Fuck you, payments. The bank's' going to come and foreclose on your family farm? Fuck you, bank. So think long and hard about all those brutal days [you spend] workin'. Fuck-you

money is the ultimate liberator.
What's under your town is fuck-you
money.

In Alberta, a world of new possibilities opened to both farmers and city people. It created new career opportunities for hundreds of thousands of them. It made it possible for Edmonton to acquire a national stage for live theatre as well as a franchise in the National Hockey League. And it made it possible for Albertans to think of themselves differently in ways big and small.

Thanks to oil, Alberta had *arrived*.

If you looked across the world, you could see that oil was sometimes no great blessing. If you lived in the Middle East, Venezuela, or Nigeria—or the swamplands of Louisiana—you could see oil contributing to corruption, autocracy, and environmental degradation. But in Alberta, and later Saskatchewan and the Maritimes, political leaders and officials ensured that oil development was well regulated and well planned and that the industry paid a fair rent for the people's resources. They also ensured that oil revenues were ploughed back into schools, universities, hospitals, and roads. People could see that oil was laying down a path to a better future, a path that could accommodate a huge number of travellers.

There was also another point. Neither in Alberta nor Saskatchewan did oil undermine social peace by creating a new class of uber-rich. Both provinces saw a vast expansion of the middle class. Leaders like Ernest Manning and the first Conservative premier, Peter Lougheed, managed oil development to ensure large government revenues and the investment of those revenues in first-class education, health care, and social services.

Sure, the day would come that some people, even in Alberta, would see oil differently and would link it to horrifying visions of planetary climate catastrophe. But that was in the future.

---

[1] See Vaclav Smil, *Power Density: A Key to Understanding Energy Sources and Uses* (Cambridge: The MIT Press, 2015).

[2] For the text of the *Regina Manifesto* see https://canadiandimension.com/articles/view/ the-regina-manifesto-1933-co-operative-commonwealth-federation-programme-fu.

PART ONE: BIRTH

1. A TRIP THROUGH TIME

Imagine boarding an aircraft that can move instantly through space *and* time. It takes us back to the Cretaceous about 100 million years ago, and we fly over the northwestern portion of North America—over what will become, in a 100 million years, Canada's prairie provinces. The oil sands are about to be born.

Below us the climate of the Northern Hemisphere is warm. An endless inland sea stretches from the Gulf of Mexico to the Arctic. Its waters are warm and shallow (as oceans go); they average 600 feet in depth. On its shores, there are vast stretches of dense jungle. Dinosaurs rustle the vegetation. Pterosaurs swoop through the humid skies.

We descend and hover close to the water. It's aswarm with dense clouds of microscopic plankton. We open a porthole and scoop up a cupful of sea water; it's full of tiny plants called coccolithophores. The biologist in the next seat says, "You can see its blooms from outer space."

We hover over the water, and then our pilot engages the time drive. The world outside our window spins by at a blur, then slows and stops; several million years have gone by. Now in the waters below, millions of generations of the coccolithophores have died and sunk to the bottom of the sea along with the bodies of countless other animals and plants. They form a dense carpet of carboniferous corpses entombed under a thickening cascade of particles of clay and sand.

The time drive speeds up; Earth ages and changes. The seas dry, leaving that marine graveyard buried under the dried-out clay of the sea bottom. The sea bottom compresses and becomes shale; the calcium of the coccolithophores becomes chalk and limestone. The graveyard groans with the weight of the overlaying rock, which compresses and heats the rocks, transforming that dead organic matter into coal, natural gas, and petroleum.

The same process is happening in many of the inland seas of what will become the Persian Gulf, Venezuela, and the Gulf of Mexico. But now, in what will become Alberta's Western Canada Sedimentary Basin, the remains of this Great Dying ever so slowly become one of the Earth's great storehouses of hydrocarbon wealth: the oil sands.

Further eons pass. Ever so slowly, the Earth's crust changes. In western North America, a collision of continental plates begins to push up the Rocky Mountains. The inland sea dries, and the Earth's climate moves through its lengthy cycles. Rivers begin to drain eastward and northward off the Rockies and commence their slow work of cutting channels through the sand and gravel.

The Rockies rise higher, overriding the sedimentary rocks on what used to be the bottom of the inland sea. In what is now southwestern Alberta, the weight of the mountains drives the sediments deeper and deeper; the sedimentary layers tilt, rising toward the northeast, where they butt up against the granite bedrock of the Canadian Shield. Deep below the ground—thousands of feet deep— the remains of the ancient coccolithophores and animals become a carbon-rich sedimentary mudstone. Under heat and pressure, the molecules within it are transformed into primordial petroleum, natural gas, and even coal. Some of it is light oil, free-flowing and rich with benzene, almost natural gas itself. This is much like the oil found in Turner Valley in 1914. In other places, it becomes heavier, denser, and more carbon-rich, which we now know as the petroleum that spewed into the air during the early days of oil drilling.

Because it's lighter than the water in the deep pores of the subsurface, it begins to soak through the overlying rock and sand deposits, creeping upward toward the northeast. In time, it'll soak into the sand beds left behind by the glaciers and primordial rivers.

Still, nature isn't finished with the petroleum under what will become Alberta. Bacterial life has been found at depths of 5 kilometres in the continents and 10 kilometres below the sea bottom. In a process not fully understood even today, geologists theorize that some of it was attacked by chemicals, oxygen, and bacteria that began to digest its lighter components, like benzene and kerosene, and leave behind its heavier, carbon-rich liquids. This sticky, heavy, barely liquid petroleum is becoming what we today call "bitumen"—the oil in oil sands.

In northeastern Alberta, the bitumen is below ground everywhere, exposed at the surface in some places, and buried a kilometre down in others.

And this isn't unique to Alberta. Remember those ancient inland seas in other parts of the world? They left behind something like bitumen in the Tigris and Euphrates valleys, in Syria, and in Egypt. Ancient peoples used it to preserve mummies and waterproof reed boats. Who knows? Maybe one of them carried the infant Prince Moses.

Now our craft hovers again. We open the porthole. Lean out and dip your fingertip into the bitumen—go ahead, it won't hurt you. Just try not to get it on your clothes. Study it carefully.

Millions of years after its birth, it'll tax the minds of everyone who encounters it. In summer, it'll stick to everything. Along the riverbanks, it'll be exposed in outcrops; on warm summer days, it'll ooze and drip into the water. In the 40-below winter, it'll harden into a gritty rock that'll grind down the hardest steel.

Over millions of years, Earth's climate has changed many times. In the last 650,000 years of our planet's life, there have been no fewer than seven cycles of glacial advance and retreat, most of them attributed to small variations in Earth's orbit that alter the amount of solar energy our planet receives.[1]

Around 15,000 years ago, the last ice age glaciers began their retreat, and the planet saw the beginning of the modern climate era—and of human civilization. Geologists say that the last ice age lowered sea levels so much that a wide land bridge opened between Siberia and Alaska; Siberian peoples migrated across this bridge or in boats along its shores and began to migrate southward through the Americas. Scholarly opinion is divided on whether the First Peoples came down a continental ice-free corridor through Alaska and Western Canada (on the eastern slope of the Rockies through what became Alberta) or down the ice-free shores of the Pacific—or both. [2]

For many hundreds of centuries, when there were no human beings present to see it, melted bitumen dripped from the northern riverbanks in the summer sun. Somewhere around 15,000 years ago, on a warming Earth, the First Peoples entered the Americas.

The land corridor through which they came might've been warming, but in all likelihood, it was still the sub-Arctic, a difficult land of boreal forest and muskeg, possibly inhabited for a period by animals that had survived since prehistoric times, creatures like woolly mastodons, giant bison, and maybe even a sabre-toothed tiger or two. The people hunted, fished, and gathered native plants and medicines. And they survived. Thousands of years later, when European traders and explorers arrived, they learned to trade the skins of trapped animals for the white man's goods.

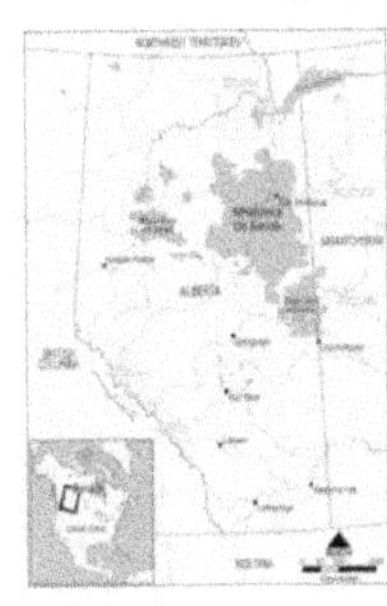

Athabasca Oil Sands. (Photo credit: Norman Einstein.)

The First Peoples had long noticed that bitumen seeped along the Athabasca and Clearwater Rivers and tapped the sticky stuff to waterproof their birchbark canoes. In 1719, a Cree man named Wa-Pa-Su brought a sample of oil sands to Henry Kelsey, a Hudson's Bay Company trader who commented on the fact in his journals.

Sixty years later, fur trader Peter Pond paddled down the Clearwater River to the Athabasca, saw the bitumen deposits, and wrote of "springs of bitumen that flow along the ground." In 1787, fur trader and explorer Alexander MacKenzie, on his way to the Arctic Ocean, saw the Athabasca oil sands and commented, "at about 24 miles from the fork (of the Athabasca and Clearwater Rivers) are some bituminous fountains into which a pole of 20 feet long may be inserted without the least resistance."

For the next century explorers, adventurers, entrepreneurs, and scientists would poke, prod, play with, and debate the potential of this vast, mysterious buried blanket of sand, clay, and bitumen, grappling with how to make it useful and profitable.[3]

Our craft has now reached the end of the 19th century, the morning of the Time of the Tar Sands.

2. THE PIONEERS

In August 2021, a piece of digital artwork entitled Everydays: The First 5,000 Days, created by Mike Winkelmann, otherwise known as Beeple, was sold at auction by Christie's for $69.3 million. It took its place alongside paintings by artists like Picasso and Monet as one of the most expensive works of art sold at auction that year.

Whether we're talking about crystals of carbon (diamonds) or primitive tools of copper, throughout human history, value has been a measure of human wants. The oil sands of Alberta—or rather, the bitumen the sands contained—were no different.

Not until well into the 20th century were the oil sands seen as having any significant value, let alone as a possible source of prosperity for millions of people. A century after *that*, the oil sands came to be seen—by many people, at least—as a threat to the world's climate. During all those centuries, the bitumen didn't change. What changed was the value (or threat) that people attributed to it. Considered in isolation of human wants and needs, things don't have any inherent value.

To the Indigenous peoples of Northern Canada and the early European explorers and fur traders, bitumen was a curiosity that had value chiefly as a waterproofing for birchbark canoes.

Although it is hard to imagine today, until the middle of the 19th century, petroleum had little value. Most of the world's ships were propelled by wind. The industrial world ran on coal. When the world's first production oil well was completed near Sarnia, Ontario, in 1858, its owners were only able to sell the light petroleum in bottles as a patent medicine; later, they sold it in pails as lamp oil.

One of the great ironies of petroleum is that its value as a lamp oil played a minor role in saving the Earth's whales, which at that point were being slaughtered almost to extinction by whalers seeking whale oil for lamps.

Society didn't place a high value on petroleum until the invention of the internal combustion engine, the mass-produced automobile that required gasoline fuel, and the oil-refining technology that made it possible to produce an increased portion of gasoline from a barrel of oil. By the time of Canada's founding in 1867, the country's miniscule oil industry, which existed chiefly in southern Ontario, was producing only tiny quantities of petroleum and selling it for about 50 cents a pail.

By 1899, the price of oil was rising rapidly, and companies like Standard Oil (of John D. Rockefeller fame) were beginning to search the globe for conventional petroleum—the kind that could be easily produced by drilling. But they weren't interested in a few springs in southern Ontario, let alone the frozen muskeg and boreal forests of northern Alberta. They were hunting major petroleum deposits in places like Iraq, Venezuela, Texas, and California.

And so it was that the oil sands chiefly attracted the interest of Earth scientists and a scattering of frontier entrepreneurs.

The Scientists

One of the ironies of oil sands history is that the first promoters of development were federal civil servants. Until 1930, when the ownership of natural resources was transferred from the federal government to the provinces, underground resources like the oil sands belonged to the Government of Canada (to say that petroleum in the ground "belongs" to someone makes as much sense as to say that the St. Lawrence River "belongs" to the people of Ontario, Quebec, or, for that matter, New York). And the first initiators of development were Earth scientists employed by the Geological Survey of Canada (GSC).

In 1893, Parliament appropriated the vast

sum of $7,000 to finance a proper investigation of Alberta's oil sands deposits both shallow and deeply buried. The GSC's Earth scientists determined that oil sands lay beneath (and sometimes were exposed at the surface of) some 70,000 square kilometres of northeastern Alberta. They believed that the oil sands constituted (as they still do today) one of the Earth's largest petroleum deposits.

In the 1890s, there wasn't much of a market for the oil of the oil sands. The region was remote, unsettled, and a somewhat forbidding sub-Arctic wilderness lacking roads or railways. The bitumen that permeated the oil sands was a source of possible future value only because of the sheer amount of it.

The GSC's investigators hoped that there'd be pools of petroleum at the base of the oil sands deposits and that these could be tapped by conventional oil drilling. The GSC wasn't alone in this thinking. Most early promoters of oil sands development imagined that riverbank seepages of bitumen indicated immense pools of petroleum deeper down, pools that could be drilled. They were wrong.

In 1897, at a site called Pelican Portage on the Athabasca River and at a depth of 235 metres, a GSC drilling crew accidentally hit a large pocket of highly pressurized natural gas. The gas blew the top off the drilling rig and rained down walnut-sized missiles of iron pyrite that endangered the workers. The roar of the stuff could be heard for more than 5 kilometres.

Several unsuccessful attempts were made to shut in the well, after which it was simply abandoned. It wasn't capped until 1918. Over that period, it spewed a quarter of a million cubic metres of natural gas into the atmosphere every day—an early contribution oil sands contribution to global warming.

Sidney Ells

In 1913, Sidney C. Ells, a McGill-educated mining engineer and GSC field officer was tasked with undertaking a complete survey of the oil sands to evaluate their commercial potential. Ells was one part man of science and another natural resource enthusiast. He left a large footprint in the oil sands' history.

Ells wasn't satisfied with exploring the sands; he spent 30 years of his life first drilling for petroleum and then digging up the shallow sands and testing ways to treat them with hot water to extract the bitumen. Ells later recounted that he became "so enthralled with the possibilities of the oil sands that [he] preferred resigning [his] position rather than being deprived of making an investigation."

Left, Sidney Ells in the field (circa 1920). (Photo courtesy of the Library and Archives Canada, PA-014454.)
Right, early barge haulers on the Athabasca. (Photo courtesy of the Glenbow Archives, NA-711-187.)

Ells was a man robust enough to pull barges upstream through the Grand Rapids cataract of the Athabasca River and carry backpacks of oil sands 250 miles through the bush to Edmonton. He was also an intensely focussed and dedicated investigator. He had to be. The oil sands didn't give up their secrets easily, and over the course of his 32-year career, most of the world took little interest in his findings. The unwillingness of others to embrace his hypotheses undoubtedly contributed to his cantankerousness.

At the beginning of his work, over the course of three months, Ells surveyed 298 kilometres (185 miles) of the Athabasca River's banks and recorded observations of 257 outcrops of oil sands. Using a hand augur, he drew more than 200 core samples from depths of 1.5 to 5.2 metres (5 to 17 feet). He supervised the loading of nine tonnes of such samples into barrels and, with a crew of Metis and Dene rivermen, helped pull the samples upriver before loading them onto wagons for overland transportation to Edmonton. At the end of Ells's initial assignment, he wrote a report that definitively described the dispersal of bitumen throughout a sand layer and debunked the myth that there were pools of liquid petroleum.

In his hunt for practical applications for the oil sands' bitumen, Ells conceived of road paving material. In 1915, he used oil sands and asphalt to lay three road surfaces on sections of 82nd Street in Edmonton. Ten years later, a report by a city engineer stated that the surface remained in excellent condition. McMurray asphalt also saw use on the grounds of the Alberta legislature, on the highway in Jasper Park, and elsewhere in the province.

Ells's paving experiment in north Edmonton. (Photo courtesy Provincial Archives of Alberta, A3399.)

Ells spent years studying hot water separation

at the Mellon Institute in Pittsburgh and came away convinced that it was a commercially viable means of processing oilsands bitumen. He was the first to realize that while there could be commercial applications for bitumen, development of the oil sands would have to address the economics of mining. He campaigned tirelessly for oil sands development, and his work helped inform several experimental oil sands recovery plants between 1920 and the early 1940s.

For years, Ells actively promoted federal investment in the oil sands and prompted Ottawa to back private developers willing to develop the oil sands. He helped establish American mining engineer Max Ball's Abasand Oils plant on the Horse River in 1936 and became lifelong friends with Ball. Years later, in 1950, Ball said, "For 35 years, in the face of indifference and skepticism, he has been the courageous and unremitting advocate of the value and importance of the bituminous sand deposits." It wouldn't be unfair to describe him as the father of the oil sands.

The Quiet Scientist

Ells's work on hot water extraction helped initiate four decades of laboratory and pilot plant process studies by Dr. Karl Clark, a professor at the University of Alberta and the Alberta Research Council in Edmonton.

Clark (1888–1966) was a quiet, serious, applied scientist and dedicated researcher. He didn't invent the idea of using hot water to extract bitumen from oil sand, but he tested and systematized it over the course of 40 years, building a series of experimental test

Clark at his University of Alberta laboratory. (Photo courtesy of the Glenbow Museum Archives, ND-3-4596a.)

beds and beginning with a small one in the basement of the power plant at the University of Alberta in Edmonton. He had to shovel the oil sand into the apparatus through a basement window.

Between 1923 and 1925, Clark tried different approaches to separating bitumen from different types of oil sand. He conducted laboratory tests and built model plants to test the processes he devised. The first test plant was on the University of Alberta campus in the basement of the Power Plant.

Under the patronage of Dr. Henry Marshall Tory, the university's first president, Clark worked diligently to develop and improve the hot water extraction process, determining how to make it work effectively on various types and grades of oil sand, until it became the base case for producing bitumen. In a major way, he laid the scientific foundation for extraction of bitumen from the shallow oil sands.

Under the auspices of Clark's work, the Alberta Research Council set up two pilot plants in Edmonton and a third at the Clearwater River. In 1930, Clark's Fort McMurray plant actually used the process to produce three carloads of oil.

As later oil companies were to discover, not every batch of oil sand processed the same. Two important variables were the amount of clay (generally called "fines") in the sand and the length of time the batch had dried in the open air. Freshly dug oil sand contained a microscopic layer of water that separated the grains from the bitumen; in the hot water process, this water layer caused the sand to separate cleanly from the oil and precipitate to the bottom of a separation vessel.

In 1927, Clark reported that "it is now practical to regard the bitumen content of the bituminous sands as a crude oil and therefore a potential motor fuel. . . . General conditions in the oil industry are not yet propitious for the development of the Alberta bituminous sands for manufacture of gasoline. But the recent improvements of the cracking process have opened up an almost unlimited outlet for separated bitumen which will be taken advantage of when the demand for crude oil becomes more keen and its price commences to rise."

Throughout his career, Clark knew all the early would-be developers of the oil sands.

Sometimes, he advised them or attempted to; other times, he competed with them. He longed to be respected as the ultimate authority on hot water extraction, and he undoubtedly deserved to be. His published studies made hot water extraction scientifically respectable and drew a degree of interest from the international scientific community and, eventually, the major oil companies, even if they weren't yet ready to invest serious money in the venture.

The Entrepreneurs

In the 1890s, demand for petroleum began to increase and quickly outran Canada's miniscule oil production in Ontario. New sources of oil were needed. Individual investors—some in Edmonton and Calgary, some in Eastern Canada, some in the United Kingdom—floated companies to explore and, hopefully, produce oil from the oil sands. Since natural resources "belonged" to the federal government until 1930, they sought federal leases and promised to send drilling crews north on the Athabasca River. Small, technically unsophisticated companies came and went, including The Athabasca Oil & Asphalt Company, the McMurray Asphaltum & Oil Limited (head office in Petrolia, Ontario), and Northern Alberta Exploration Company Ltd.

A few of these companies managed to attempt drilling operations after using flat-bottomed scows to negotiate the rapids above

Fort McMurray. On average, a trip took eight brutally hard days on the river. The first oil sands developers to actually produce any oil from the sands made their debut between the turn of the century and the start of the Second World War.

Two of these men—Alfred Hammerstein (1870–1941) and Robert Cosmas Fitzsimmons (1881–1971)—were frontier entrepreneurs with big plans, little money, and only a rudimentary understanding of what oil sand was or how to separate bitumen from it. Both went broke. Max Ball was the third, and he was more of an authentic oilman with scientific training and a modicum of financial support. His projects made some progress before burning down; ultimately, he also went home unsuccessful.

Hammerstein, Fitzsimmons, and Ball deserve much credit for tackling such a monumental challenge in a place as remote, intimidating, and inaccessible as northeastern Alberta. It was the last part of the province to be developed for agriculture or anything else. Prior to the Second World War, most of it was bush with no real roads. Its only significant settlement, Fort McMurray, had a population of just a few hundred souls, with few skilled workers other than a small number of Metis men who worked as captains on Athabasca riverboats. The main street of Fort McMurray was a dirt road. In winter, it was mainly travelled by dogsleds. Before 1925, when the rickety Northern Alberta

Railway reached Waterways south of McMurray, the only way to get equipment in to the oil sands region was to ship it down the Athabasca River, a task that required manually hauling barges through the formidable Grand Rapids, a cataract south of Fort McMurray.

It's worth noting that until halfway through the 20th century, Big Oil had little interest in the oil sands. The North American and European majors were focused on finding major deposits of conventional oil—the kind you drill for—in the Middle East, Venezuela, Mexico, Texas, California, and Louisiana. Not even when conventional oil and natural gas were discovered in 1914 near Turner Valley did most international oil companies show any interest in Alberta, let alone the oil sands 800 kilometres farther north—and most of them lost interest as the Turner Valley oil field declined.

One major offered an exception: Royal Dutch Shell. In November 1918, Shell asked the federal government to lease the company a concession of 250,000 square miles—later amended to 328,000 square miles—of northern Alberta and parts of Saskatchewan and the Northwest Territories for oil exploration and production. This audacious proposal was widely opposed in Western Canada, and the prime minister of the day, Arthur Meighen, dodged it until it died a natural death.

The first generation of would-be oil sands

developers had only one thing in common: a belief that the oil sands had vast potential and that they could build a profitable business exploiting it. *How*, exactly, the first two men weren't quite sure.

Hammerstein—or, as he liked to call himself, *Von* Hammerstein—was certainly an entrepreneur; among other things, he ran a freight hauling company and owned a small hotel at Athabasca Landing. And he did sink an exploratory oil well or two. I imagine him as a somewhat Germanic Willy Loman of *Death of A Salesman*, "a man way out there in the blue, riding on a smile and a shoeshine."

Left, Hammerstein in the field. (Photo courtesy Glenbow Archives, PA-3920-1). *Right*, Hammerstein as a hotelier. (Photo courtesy of the Canadian Petroleum Hall of Fame.)

Hammerstein said he was a German count, a claim generally discredited by historians. He arrived in the region in 1897 on his way to the

Klondike Gold Rush. Captivated by the oil sands, he decided to stay and exploit the resource. He kept at it off and on for over 40 years, taking photos with descriptive titles like *Tar Sands and Flowing Asphaltum in the Athabasca District*. He imagined that there were pools of oil at the base of the sands and that they could be tapped by drilling. He sunk a number of wells on the banks of the Athabasca and one of its tributaries, the Horse River. He was one of the first to find that the bitumen in the oil sands was thick, viscous, and couldn't be brought to the surface by conventional means.

Hammerstein's syndicate received the first (and only) clear title to oil sands lands in 1910. He was able to produce very little oil and died in 1941 in relative poverty. He was elected to the Canadian Petroleum Hall of Fame.

Fitzsimmons, the Giant of Bitumount

Hammerstein drilling rig on the bank of the Horse River south of Fort McMurray. (Photo courtesy of the Library and Archives Canada, PA-138536.)

The previously mentioned Fitzsimmons was a prototypical little guy with a vision, in well over his head. A gritty entrepreneur from Prince Edward Island, he pursued a career in real estate before

23

becoming an oil sands developer working out of Edmonton. Despite many setbacks, he stuck with his oil sands vision for more than 20 years.

In the early 1920s, he acquired oil sands rights on land about 90 kilometres north of Fort McMurray on the east bank of the Athabasca River, a location he later named Bitumount. Through his tiny company, Oil Sands Ltd., Fitzsimmons built not only the first oil sands separation plant developed by private industry but also showed it was possible to separate and refine the oil sands on a significant scale.

Bitumount, probably in the 1940s. (Photo courtesy of the Provincial Archives of Alberta, PAA-91-137-172.)

Like others, Fitzsimmons first attempted to drill for oil; like others, he found it futile. Then, between 1922 and the mid-1940s, his International Bitumen Company Limited mined shallow oil sands deposits and built a small-scale hot water bitumen-extraction plant.

Though he lacked scientific training and experience in processing oil, Fitzsimmons had no shortage of common sense and devoted many years of tireless labour and practical experimentation to figuring out how to extract bitumen and market the products of the oil sands. One year, he managed to extract and either sell or give away 307 barrels of bitumen, promising customers that it was valuable "for laying built-up roofs, processing into plastic gums, lap cement, caulking compounds, fence post preservatives, boat pitch, belt dressing, mineral rubber and skin disease medicine."

In a self-published book entitled *The Truth About Alberta Tar Sands: Why Were They Kept Out of Production* (Edmonton, 1953), Fitzsimmons reported that as early as the mid-1920s, he had shipped small quantities of bitumen to chemical and refining companies to explore its potential in everything from gasoline to consumer products. In 1931, Fitzsimmons said, his company produced 2,000 barrels of bitumen for gasoline.

But his bitumen seldom had many buyers, and he was often forced to donate it to various enterprises for testing and testimonials. By 1942, he was in desperate straits. Bitumount had opened and closed several times and never made a profit. His company owed money all over town and had been struck from the financial registry. A Montreal financier, Lloyd Champion (1904–1971), offered to invest in the company, pay off debts, and get the plant back into operation in exchange for a controlling interest.

Fitzsimmons (left) with Champion, the venture capitalist who ousted him from his own company. (Photo courtesy of the University of Alberta Archives, 83-160-113.)

Like the legendary hippo asked by a scorpion to ferry him across the Congo River, Fitzsimmons was soon to discover that some offers come at a steep, hidden price.[4] Champion, a kind of early-day vulture capitalist, manoeuvred Fitzsimmons out of any role in running the company and then decided to build a new operation at Bitumount. In 1944, he negotiated a deal with the Government of Alberta to build a new pilot separation plant from scratch. Champion knew less about oil sand than Fitzsimmons and much less than Fitzsimmons about building or operating a plant. He ignored advice from both Fitzsimmons and Clark, attempted to build a new plant from scratch, and failed. In 1948, the Alberta government cancelled Champion's lease and took control of Bitumount.

Champion didn't disappear immediately. In March 1948, he briefly appeared in a front page story in *The Edmonton Bulletin* announcing that he had enlisted Gene Tunney, former World heavyweight boxing champion, to play a part in a new oil sands company. Subsequently neither Champion, Tunney, nor their so-called company were heard from again.

As for Fitzsimmons, he finished out his days complaining—not without cause—that the International Bitumen Company was the victim not just of the Depression but of both federal and Alberta governments' short-sightedness and lack of support. He hinted darkly that "big interests" had influenced the government to ignore his entreaties and that given proper encouragement and recognition, Bitumount could've been a very viable business.

The new Alberta government-owned Bitumount plant conducted successful tests of the Clark hot water extraction process, then converted it into a pilot demonstration project for international visitors. After the Alberta government commissioned Sidney Blair to write a major report announcing that oil sands development could be financially sustainable and sponsored a big international conference to promote the idea, the Bitumount plant was finally shut down and eventually abandoned to the muskeg.

In his nomination to the Petroleum Hall of Fame, Fitzsimmons' grandniece wrote that he "helped propel the industry that is ongoing today and the industry has, in turn, stood on the shoulders of Fitzsimmons and other pioneers like him."

Ball, the American Geologist and Engineer

Left, the Abasands Plant, on the Horse River, circa 1943. (Photo courtesy of the Library and Archives Canada, PA-017285). *Right*, Max Ball. (Photo courtesy of the University of Alberta Archives, 89-120-008.)

Max Ball (1885–1954), the third and last entrepreneur to have a go at the oil sands, was an American with serious credentials as a geologist and oilman as well as connections with the US Department of the Interior. He graduated from the Colorado School of Mines with an engineer of mines degree and got his first career job with the United States Geological Survey. In 1929, Ball became aware of the potential of the Athabasca oil sands, probably through J. M. McClave, a mining engineer also based in Denver who had a longstanding interest in the oil sands. After forming a syndicate with McClave and Basil O. Jones, an investor specializing in petroleum companies, Ball became involved.

In February 1930, Ball, McClave, and Clark visited the Alberta Research Council Clearwater River pilot plant. Within a few months, Ball obtained Bituminous Sand Permit No. 1 from the federal government, which allowed him to set up a mining and extraction operation on federal land along the Horse River near Fort McMurray. (The federal government had transferred mineral rights to the Western provinces in 1930. Ball's lease was on a slice of land that had been reserved to the federal government for future parks.)

Ball and Jones reportedly had a secret recovery method known as the McClay process, and they claimed substantial financial backing. In 1935, Abasand Oils Limited, Ball's American-backed operating company, started construction of a new plant west of Waterways. Under the agreement with the federal government, the plant was to be operational by September 1, 1936. But forest fires and failure of equipment suppliers to meet delivery dates delayed completion.

Ball inherited from Ells, Clark, and Fitzsimmons the idea that that the shallow oil sands could be mined and the bitumen extracted with hot water; in addition, he had ideas about how bitumen could be upgraded into commercial products like gasoline, diesel, and kerosene. Between the mid-1930s and early 1940s, on the very cramped Horse River site, Abasands Oil mined oil sand with a bulldozer, used hot water to extract the bitumen, and ran a primitive upgrader to upgrade it into gasoline and diesel fuel.

Abasands produced a quantity of merchantable gasoline and diesel fuel but was cursed with operating problems (including the notoriously sticky and abrasive oil sand). It burned down twice. By the end of September 1941, Abasands had produced 2,690 cubic metres of oil, but two months later, fire destroyed the plant. Rebuilt on a larger scale, it was fully operational in June 1942. In 1943, the federal government decided to aid oil sands development and took over Abasand.

C.D. Howe, the federal wartime industry czar, announced that federal "experts" were going to assess the oil sands' potential to become a major supplier of oil, though later, he stated his doubts. So-called federal assistance proved chimerical. The government's researchers concluded that the hot water process was uneconomic because of the extensive heat loss, and they proposed instead an untested cold-water process. Federal work at the Abasands plant came to an end with a disastrous fire in 1945.

The Abasand mining, extraction, and refining complex produced significant amounts of oil, but in the end, Ball was defeated by operating mishaps, fires, and erratic and unreliable support from the federal government. His place in the oil sands story is a curious one: he was the first developer to bring technical knowledge and significant financial backing to the oil sands, but ultimately, he was hardly more successful than

Fitzsimmons, who had neither.

By the end of the Second World War, the decaying physical remnants of these early commercial pilot plants could be seen in several places along the Athabasca River Valley, and the memory of their owners' efforts drifted like mist through the muskeg.

3. THE PREMIER AND THE TITAN

The Premier

For almost two centuries after the first sightings of the oil sands by European fur traders and explorers, the deposits slumbered. Earth scientists and a few frontier entrepreneurs poked and prodded the sands and even built primitive pilot plants to tap the bitumen, but by the early 1950s, serious development was still very much a one-of-these-days proposition. If the oil sands were ever to become a major source of petroleum, a catalyst was needed. When one finally came, it emerged from a surprising direction: a quiet, thin Saskatchewan farm kid who'd come to Alberta to study the Bible under an early-day radio evangelist.

Ernest C. Manning had tuned into the hour-long *Back to the Bible* radio broadcasts of William Aberhart in the late 1920s and was deeply drawn to the man and his message. He went to Calgary to study under Aberhart at the Prophetic Bible

Institute, and in the depths of the Depression, when Aberhart decided to enter provincial politics, Manning went with him. He became Aberhart's aide-de-camp and, after Social Credit was elected in 1935, a cabinet minister. When Aberhart died of cirrhosis in 1943—

Ernest Manning, circa 1943. (Photo courtesy of the Glenbow Museum NA-2922-14.)

surprisingly, since he was a teetotaler—Manning inherited the premiership. This was partly thanks to their close, almost father-son relationship and partly because Manning had fewer enemies than any other potential successor.

In those years, many people underestimated Manning. He was quiet, serious, and uncharismatic. From the time he attended the Prophetic Bible Institute until his elevation to premier, he went almost unnoticed in the glare of Aberhart's outsized personality. It took a while for people to realize that Manning's quietness concealed a formidable intelligence that was both practical and, when required, visionary. And that he was, quietly, his own man.

It may sound strange to describe a man who led Alberta for 28 years and seven elections as outside the political class. Still, it was true. Alberta's political class was overwhelmingly urban, university educated, and almost always Liberal or Conservative. Social Credit was a genteel populist party that refused to call itself a political party at all for 30 years.

Manning came to office with no experience in business and no education beyond Grade 12. He had no experience with the oil industry and no recorded interest in it. Like most Canadian politicians at the time—remember, this was 1943—his thinking was occupied by the Second World War and its issues. Alberta had a small, rapidly declining legacy oil industry rooted in the province's first oil discovery at Turner Valley. By 1942, Turner Valley's production had been cranked up to 27,000 barrels a day to support the war effort, but that pace was dropping rapidly since the reservoir had been prematurely depressurized; by the end of the war, Canada was consuming 10 times as much oil as it produced, Turner Valley was in steep decline, and the country was importing 90 percent of its oil requirements from the United States.

For 30 years, the province had encouraged the oil industry to find the next Turner Valley, but by the end of the war, prospects were bleak. The biggest explorer, Imperial Oil, which had a long history of interest in the West and North —this going back to the discovery of oil at Norman Wells, Northwest Territories, in 1920— was on the verge of writing Alberta off. Imperial had drilled 133 dry holes with little or nothing to show for it. Then, on February 13,1947, Imperial's last-ditch gamble on a deep well near

Leduc finally paid off.

Leduc #1, as the well was called, marked a turning point in Alberta's economic and social development. Leduc-Woodbine, as the oilfield came to be called, was a major discovery. There'd be more to come before long, with names like Redwater and Devon and Swan Hills and Rainbow Lake, all of them conventional oil fields and not sands.

The Leduc discovery transformed Alberta's economy; oil and gas soon supplanted farming as the primary industry, and the flow of oil royalties and exploration bonuses soon made the province one of the best-off in the country. Swaggering men (and they were men) with Texan and Oklahoman accents suddenly crowded the streets, and before long, the Petroleum Club replaced the Ranchmen's Club as the city's social centre.

Alberta's new oil discoveries allowed Canada to become self-sufficient within a decade and, ultimately, a major exporter of oil. Revenues from lease sales and oil production royalties poured into the province's coffers. Manning's government spent this income wisely, using it to fund a vast expansion of the province's post-secondary education system, social services, and health care. This creation of a "welfare state without welfare-state levels of taxation" created a broadly based prosperity that became, seemingly, a permanent feature of life in Alberta.

It's greatly to Manning's credit that while his government reaped this cascade of cash, it focused significant attention on the creation of a regulatory infrastructure to regulate the booming oil industry—and Manning kept one eye on the immense but untapped potential of the oil sands.

In 1948, he set up a board of trustees for Bitumount that included one of his most capable cabinet ministers, Nathan Tanner, along with Public Works Minister DDB MacMillan, J.L. Robinson, George Clash of the province's crude oil marketing board, and William Adkins, an engineer who used to work with Robert Cosmas Fitzsimmons. The government appropriated $500,000 to fund the construction of a new Bitumount pilot plant that would process 500 tonnes of oil sand a day. Karl Clark's help was sought to design it.

On August 26, 1949, Manning took 40 members of the Alberta legislature on a tour of the plant. Apart from the fact that there was an explosion and fire the day of the visit, it went well.

The next year, Manning asked Sidney Blair, who'd been Clark's assistant, to prepare a comprehensive report on the oil sands outlook. In May, Blair—who went on to head the engineering and construction firm Canadian Bechtel—released his *Report on the Alberta Bituminous Sands*. Blair estimated the cost of mining, separating, upgrading, and transporting oil sands petroleum to the Great Lakes would total $3.10 a barrel, compared with the prevailing oil price of $3.50. He said a 20,000-barrel-per-day project would cost $43 million to build and would generate a 5–6 percent return.

After disseminating the Blair Report through the oil industry, on September 10, 1951, Alberta sponsored a week-long international conference on oil sand development in Edmonton. Over the course of the proceedings, Tanner released a 1,000-word policy on oil sands leasing and royalties. It provided for companies to take out renewable "prospecting permits" for up to 50,000 acres. The policy statement said, "The Alberta Government is desirous of doing all that is reasonably possible to encourage the orderly development of the enormous oil sands deposits in the interest of the people of the Province and of Canada as a whole, and, further, for the security of the continent." Initial royalties would not exceed 10 percent, and lease rentals would be $1.00 an acre.

Manning hoped to start a horse race to develop the oil sands, but in the decade following Leduc, industry interest in the oil sands was more of a tortoise race.

In the months and first years following the conference, a dozen oil companies nibbled the bait and took out exploration permits. But none immediately went to the lease stage because the government stipulated that a lessee had to begin construction of a commercial plant within two years of receiving a lease and begin operating the plant within five.

No one knows what Manning's expectations were when he opened the gate to prospective oil sands developers; but he had to be patient. The idea of mining oil didn't immediately sit well with most oil companies. The culture of the industry was "explore, drill and produce." Plus, there were concerns and questions about the terms of the government's proposed oil sands royalty regime; the concern centred on the treatment of operating costs. This question was to preoccupy relations between the oil industry and the Alberta government for years.

By 1952, five companies had begun nibbling at the bait, but only one of them—Pacific Petroleums Ltd.—was a significant industry player, and none of them were in any rush to proceed beyond taking out exploration leases. It took nearly 10 years for a qualified oil sands applicant to step forward, come to an understanding with Manning, leap regulatory hurdles, and build the first commercial oil sands plant.

The initial focus of the oil industry was the so-called shallow oil sands, those that were close to the surface, with less than 100 feet of sand and gravel overburden. They'd need to be mined and treated to remove the bitumen. These shallow deposits represented only about 10 percent of the total oil sands but were, in themselves, one of the world's major oil deposits. The deeper deposits—those down to 2,500 feet below the surface—would need to be attacked with in situ technologies to separate the bitumen in place and pump it to the surface. The technology to do this wouldn't be developed until the 1980s.

Part of the reason that most of the majors had feeble interest in the oil sands was that conventional oil fields were their stock in trade. They knew how to find conventional oil reservoirs and drill them, how to refine what they pumped and sell its products. That was their business model. Producing and marketing bitumen from the oil sands required different technology, a different business model, and a completely different mindset. For one thing, oil sand had to be mined, and oil industry people didn't "get" mining. Not only didn't they get it, they didn't *like* it. In 1974, I asked Walter Prior, a greying petroleum engineer who'd worked most of his life at the Sarnia refinery, what he thought of bitumen. "I'm used to working with good petroleum," he said. "Not this shit."

As later events were to prove, the majors had a lot to learn about both bitumen and mining, especially in a tough, subarctic place like the oil sands, and learning on the job was going to be very, very expensive.

In 1953, Lloyd Champion, the Montreal

financier who'd displaced Fitzsimmons at Oil Sands Limited, created a small (mainly paper) company called Great Canadian Oil Sands Ltd (GCOS), and the firm began acquiring patents and leases in the Fort McMurray area. Whether Champion got rich when Sun Oil acquired GCOS —a question that Fitzsimmons would've found interesting—isn't known these days. The part played by Montreal investors in the creation of GCOS or its sale to Sun Oil hasn't been properly researched, at least to my knowledge. Did Champion reap a windfall for his early investment in Fitzsimmons's struggling little company? The answer's lost.

The Titan

In the early 1950s, the years immediately following his opening of the door to oil sands development, all Manning got was nibbles from minnows. Then, in the early 1950s, he spotted a big fish scouting the pool. It was the chairman of Sun Oil Inc., J. Howard Pew (1882–1971).

Pew before the US Congress. (Photo credit: Harris & Ewing.)

Pew's father, Joseph Newton Pew Sr. (1848–1912), started the Sun Oil Company in Pennsylvania before migrating to Texas, where he built a substantial independent oil company and began a long business (and religious) rivalry with the Standard Oil Company controlled by the Rockefeller family. In 1886, at the age of 25, after striking oil in Beaumont, Texas, Pew Sr. began building a refinery on a barren 82-acre site along the Delaware River in Philadelphia's Marcus Hook district.

Pew's oldest son, J. Howard, as he was known, studied petroleum engineering and took over management of the family company at age 30. He and his brother introduced many production, processing, and marketing initiatives, and during the First World War, they invested in the transportation of crude oil by ocean tankers, a lucrative business that Sun built on during the Second World War. Sun was the major shipper of aviation gasoline to Allied forces during the war. By the end of the fighting, the Pew concern was the largest private shipbuilder in America.

J. Howard's experience as an oil shipper strongly stimulated his appreciation of the strategic importance of oil and the imperative of North American oil independence (since a number of his tankers were sunk by German U-boats).

He was highly intelligent man—he graduated from MIT with an engineering degree at 18 —and he was very politically conservative; he and his family were major financial contributors to the right wing of the Republican Party. For both religious and ideological reasons, the Pews were serious enemies of the Rockefellers and the moderate Republicanism they represented.

One of the richest political ironies of recent times is that the Pew family trusts—funded by oil money and created to support Christian and right-wing causes—evolved into a funding powerhouse of left-wing and environmentalist causes, most notably opposition to the oil sands. In 2005, the Pew Environmental Group had an annual budget of $70 million principally focused on climate change and ocean policy. Within a few years, the Pew Environmental Group, the Moore Foundation, the Hewlett Foundation, and the Packard Foundation together spent $57 million on programs to oppose oil sands development.[5] In all likelihood, this exceeded the fossil fuel companies' oil sands–related public relations budgets by an order of magnitude.

From Glenmede, the family mansion in the very wealthy Main Line suburbs of Philadelphia, J. Howard carried on serious religious work (he was an early sponsor of Billy Graham) as well as his business and political work. In 1957, *Fortune* listed J. Howard and his brother as among the country's wealthiest people alongside J. Paul Getty, Doris Duke, Mrs. Edsel Ford, Joseph P. Kennedy, and several members of the Rockefeller, Mellon, and DuPont families.

Sometime in the 1950s, J. Howard's interest in continental oil security saw his attention turn to Athabasca oil sands development. Whether he and Manning met in the course of their respective religious work or through their mutual interest in the oil business wasn't recorded.

Nevertheless, Manning needed to attract an investor with deep pockets who might be willing and able to build the first oil sands megaproject despite its serious technical and economic risks. (GCOS would eventually discover the seriousness of the risks when its oil sands project experienced huge cost overruns, chronic operating problems, and plant shutdowns.) In the early to mid-1950s, none of the major oil companies had any serious interest in the region. Sun Oil—a large and so-called independent oil company because it's not part of the Seven Sisters—was a major enterprise with significant resources and its own way of looking at the industry. Sun was to become the first significant firm to express an interest.

How the Deal Was Done

How or when, exactly, Manning and J. Howard agreed that Sun Oil would build the first industrial-scale (or "serious") oil sands plant wasn't recorded, but some details are known.

J. Howard was highly intelligent and widely read; he was known to appreciate the potential of the oil sands (as well as the potential of other petroleum sources like the Colorado Oil Shales). It seems likely his reading would've included the Blair Report. As an early and strong advocate for continental energy security, he would've seen the vast reserves of the oil sands as secure for decades.

J. Howard hated big government—some would say *all* government—and it took a lot to get him to make a deal with a premier, but Manning was a shrewd thinker and a conservative of the kind J. Howard liked.

Both J. Howard and Manning were highly active lay Christians. True, they came from different traditions (Manning a Baptist, Pew a

Presbyterian); beyond that, they saw in each other a deep commitment to the importance of Christian faith. J. Howard read the Bible daily and was an early discoverer of Norman Vincent Peale and Graham. At one point, J. Howard dressed down to attend one of Graham's early sawdust-floor revival tent meetings and was an early financial supporter of the Graham Crusades. In the 1950s, J. Howard invited Graham to stay with him at the Jasper Park Lodge in Alberta, and on one occasion, he and Manning conferred with Graham.

In a communication with me, Preston Manning, son of Ernest, recalled a meeting at Jasper Park Lodge between his father, J. Howard, and Graham during which they talked about the importance of strengthening the spiritual side of North American culture. The conversation established common interests and built rapport.

> Once construction was underway, my father and J. Howard would meet periodically to review progress. Sometimes, they'd meet at the Point Cabin at the Jasper Park Lodge, which [J. Howard] rented in the summer and from which he would fly back and forth to Fort McMurray. On a few occasions, as part of my political education, I was privileged to go along to those meetings. It was at those

> meetings that I first heard the phrases "North American energy security" and "continental energy security" spoken by J Howard Pew—concepts which didn't even enter the vocabulary of most political people until after the OPEC oil crisis of 1973. [J. Howard] was a visionary with a strategic view of the role petroleum resources play in the world—more than 20 years ahead of his time on that subject.

A few writers have imputed significance to the Manning-J. Howard-Graham spiritual relationship, but it's hard to imagine that it was anything more than a catalyst for their eventual agreement to go forward with GCOS.

One of the issues they had to confront was whether the Alberta government would set aside a market niche for oil sands petroleum as opposed to conventional oil. This was not a minor matter. From the 1950s and 1960s forward—and still today—there was great reluctance in Eastern Canada to buy Alberta oil (which was more expensive than imported oil from Venezuela and the Middle East). Oil sands petroleum promised to cost even more than Alberta conventional crude. Who would buy oil from the sands?

This question boiled up again during the pipeline debates of 2019. Quebec politicians got to enjoy the comfortable position of opposing what they called "oil sands pipelines" across the la belle province both because the pipelines contributed to climate change and because they'd require Quebeckers to pay more for oil from the oil sands than for oil imported from the United States, much of it fracked, or oil imported from the Middle East. This is an indictment of Quebec politicians, not ordinary Quebecers. In public opinion polls 2020, a large majority of Quebeckers said they preferred buying oil from the oil sands than from foreign sources.

Getting J. Howard to a yes required the building of personal trust. Said Preston Manning,

> In the end, after all the technical reports, feasibility studies, and financial reviews, it came down to this: my father saying to Pew, "I'm prepared to believe that you can build this plant when nobody else has been able to do so thus far"; and Pew saying to my father, "And I'm prepared to believe that you will guarantee a portion of the market for the output of this plant without which it cannot be financed." And so they shook on it, and the plant was a go."

Except it wasn't. Not yet. The premier made it clear that Sun Oil would have to make a proper application to the Oil and Gas Conservation Board and meet whatever conditions the board laid down. He and the Alberta political class knew that Big Oil, through political donations, had corrupted many American politicians; Manning wanted none of that in Alberta. The Oil and Gas Conservation Board had been created in the 1930s to impose some order on the oil industry's wild west drilling practices, and Manning insisted on putting as much distance as possible between that strong industry regulator and the political arm of the province, specifically his ministerial cabinet. J. Howard would have to make his case to the board.

As a side deal, Manning and J. Howard agreed that Sun would make it possible for ordinary Albertans to own a portion of the GCOS plant by offering 6 percent convertible debentures of which every Albertan could purchase up to a maximum of $12,500 (more than $100,000 in current dollars). If Sun's public shares grew in value, people could convert the debentures to Sun stock; if they didn't, people could enjoy a good rate of return on their investment and eventually get their capital back. This was an innovative step in the direction of partial public ownership of an oil sands project. It was an echo of the famous citizen share issue of Alberta Gas Trunk Line in the mid-1950s, an offer that thousands of Albertans subscribed to at an initial share price of $5 per share. Many of those investors would reap a 500 percent return.

But no one could see the future, and initially, old-timers in Fort McMurray were understandably skeptical that GCOS would ever make good on the promise of oil sands development. "Come on, sonny," they told an early visitor with news of the announcement. "We've been hearing that story since Peter Pond came through in 1787."

And yet, both men kept their word. In an episode that became an oil sands legend, J. Howard met with the Sun Oil Board of Directors and announced that if they wouldn't authorize Sun investing in GCOS, he'd do so himself, presumably by selling some or all his shares in the company. J. Howard was the controlling shareholder of the company; he was also, well . . . he was J. Howard Pew. So the board of directors caved in. Several of Sun's shareholders were openly critical of the decision and said so in no uncertain terms at Sun's Annual General Meeting. One of them referred to GCOS as a "white moose" (a Canadian white elephant, presumably). As for Manning, he kept his word to support the project as it moved through hearings at the Oil and Gas Conservation Board.

In 1962, GCOS received approval from the Oil and Gas Conservation Board to build and operate a 31,500-barrel-a-day plant at Tar Island near Fort McMurray. There were two other applicants: Cities Service Athabasca Inc., predecessor of

Syncrude, and Shell Canada. Cities' application was for an 80,000-barrel-a-day mining plant north of Fort McMurray; Shell proposed a 100,000-barrel-a-day in situ plant at the Peace River deposit using steam drive oil recovery technology, something very little tested at the time.

In due course, the Oil and Gas Conservation Board—led by Chair George Govier, a short, dapper University of Alberta engineering professor—weighed the merits of GCOS against two rival applications from Cities (predecessor of Syncrude) and Shell Oil Canada. The board opted for GCOS. There were two reasons: first, GCOS was the smaller of the two proposed projects and would carry less risk of endangering the conventional industry's share of markets; second, Sun committed to take a share of GCOS's production for its Marcus Hook refinery in Pennsylvania, a market not previously open to Alberta petroleum.

The Oil and Gas Conservation Board wanted to protect the province's conventional oil producers from major competition from oil sands developers, so it deferred both the Cities and Shell applications for five years. In 1968, the province changed its policy on bitumen's share of the oil market by raising the allowable number of barrels of bitumen produced from the oil sands. This opened the door a crack for Syncrude, but the change came too late for Shell, which withdrew its proposal for the Peace River project that same year.

Thinking back on it many years later, Preston Manning mused about J. Howard's motivation for wanting to build GCOS:

> If you ask that question of young people or media folk today, the most frequent answer is "to make money." But that's nonsense . . . Pew was head of one of the wealthiest families in the United States. And as he was approaching 80 years of age, he had already willed most of his personal fortune to a charitable trust and charitable projects. So why [did he do it]? He explained to my father that the Pew family had made much of its early fortune through the Sun shipbuilding company. That company built a large number of the oil tankers which supplied the allied forces in both World War I and World War II. Pew knew how many of those tankers had been sunk by enemy submarines and destroyers and he had become convinced that the United States—indeed North America—was vulnerable from a security standpoint because of its dependence on offshore oil . . . It took a terrible war to get business men like J. Howard Pew and politicians

like my father and C.D. Howe in Ottawa to think strategically about petroleum resources—its role not just in energizing our cars and heating our homes but in shaping security, trade, and dependency relations between economies and nations.[6]

A Bridge Too Short

Then the road got bumpy indeed. GCOS was supposed to be a proof-of-concept project that would entice developers into a bright future. Instead, for a number of years it appeared to lead them into a swamp.

GCOS hired Bechtel Corporation of San Francisco to be the plant's managing contractor. Bechtel built a scale model of the plant in Edmonton, a pilot plant at Tar Island to test out mining technology, and a construction camp that held more than 1,000 workers. Bechtel sent out a call to international suppliers of everything from pipe and electrical components to huge steel vessels. And GCOS's costs began to mount steadily: from an initial estimate of $122 million to $190 million and finally, by plant completion, to $250 million ($1.6 billion in 2020 dollars)—at the time, an appalling 50 percent inflation rate.

It took 22 years after GCOS's start for the company to earn back its cost of capital, and even then, only when the federal government permitted it to charge the world price for its oil.

GCOS was certainly a landmark, but a launching pad for the oil sands industry it was not (unless we mean the kind of launch pads in the early days of the space program, the ones on which rockets blew up).

GCOS built the world's first commercial-scale oil sands plant, and it pioneered technology for bitumen extraction and upgrading. It also spawned its share of oil sands legends. It was said, for instance, that the plant's German-built, electrically operated bucket-wheel excavators drew so much power from the Alberta electrical grid that under certain conditions, they had the potential "to blink the lights in Edmonton and Calgary." Years later, Syncrude utility engineers generated an even better legend: if Syncrude's four giant draglines all pulled at the same time, they "would dim the lights as far south as San Diego."

From the day of its official opening in 1967, GCOS stumbled through the tough realities of oil sands development. J. Howard was visionary, but Sun Oil's operating management lacked the skill to commercialize his vision. Both Sun and Bechtel made major mistakes. For the first years of its life, GCOS was a trainwreck.

In 1968, Sun sent in a man to fix the problems. His name was Harold Page, and he was a tough, profane executive from Dow Chemical in Europe. Page later said, "It was obvious that even from a preliminary discussion, they were in deep trouble. The whole plant wasn't running, and they couldn't get it started."

Page's recounting of Sun's failures was long, detailed, and bitter. The company's shortcomings included intellectual arrogance and the failure to hire virtually any engineers with manufacturing or operating plant experience. Page said that Sun Oil was like Dow Chemical: "a big American company that knew all the answers." Bechtel came in for its share of blunders, chief among which was "designing the plant in California for California climatic conditions."[7]

According to Page, Sun Oil failed to include experienced operating and maintenance people in the design team, failed to build a community in Fort McMurray that could provide family and community support for skilled employees and their families, and especially failed to anticipate the number of skilled people that'd be required to operate a highly technological operation in a very tough physical environment.

In an interview years later, Page pulled no punches:

> See, Bechtel and Sun Oil declared GCOS completed and ready to operate in the fall of 1967 . . . But it just didn't. It just bloody well wouldn't start. This was before my time, but I knew from talking to people once I got up there. In the

winter of 67–68, they had created the biggest ice sculpture that had ever been made. The steam plant was designed by Bechtel in California; it did not take into account the fact that the extraction plant, which required steam to mix the oil sands [and] water and caustic to extract the bitumen, demanded more steam when the incoming grade of bitumen was low. And the steam plant was not designed to accommodate those variations. And as a result of it, their condensate tank overflowed at -30F to -40F, and it just froze up the whole thing.

GCOS, Page added, was a mining company, and Sun, who knew a lot about oil refining, knew nothing about mining.

The oil sand mining industry . . . involves surface mining . . . more material than the rest of surface mines put together in the rest of Canada. And much more technologically demanding, and that, in turn, requires a very large manpower force. Not only to operate it but also to maintain very complicated equipment in winter. The mining aspect of the thing was a real dilemma . . . When you expose the [mine] face in the summer, the bitumen . . . runs like syrup, and the mine face is very unstable. In the winter, it turns hard as concrete. Bechtel, of course, are recognized worldwide as mining experts, and the plan they devised was to remove the overburden with scrapers, and that worked all right. And to mine the oil sand face with bucket-wheel excavators, which they got from Germany. Like most things in Germany, they are very, very well advanced. They have used bucket-wheels for their coal industry, and they manufacture and sell them all worldwide. And they are very efficient, except in the winter in the oil sands, when it turns to concrete. Plus, the fact that the steel that they use in the bucket-wheels that were purchased for GCOS had a critical temperature of—I think it was in the region of -30 to -40F. Anything below that, you ran the risk of the steel cracking. Well, the winter of 68–69 . . . we were measuring temperatures on the mine face of -52F.

In those terrible first winters, GCOS's bucket-wheel excavators frequently shut down after their teeth broke on the frozen oil sands. Huge frozen chunks of oil sand passed through the excavators and tumbled onto the conveyor belts, breaking them. This brought the processing of oil sand to a standstill. Bechtel, Page added, had no solution for the freeze-up of the oil sand mining face. That had to come from one of Page's hires, a veteran Canadian mining engineer who suggested breaking it up it with high explosives. Page said,

> I thought, "Oh, Jesus." Because, the mine face at that time was only, oh hell, 300 yards or so from the upgrader. In any event, [we] decided, "Well, what the hell. This has got to be done." So we brought in a drill, brought in ammonium nitrate, put a fuel line on it, and blasted the [mine] face. That's the only way that we were able to mine the oil sand face in the winter of 1968–1969. The only way it could be done. But that [was a] local solution. [The problem of winter temperatures] should have been looked at it [by Sun Oil and Bechtel when the plant was being designed]. It was obvious before the money was ever committed that that stuff was going to freeze over winter.

GCOS was supposed to show the world that an oil sands mining plant could be designed, built, and operated successfully. What it showed instead was that Sun Oil's project management capacity had been overwhelmed by the challenges and that Bechtel had badly stumbled in trying to design an oil sands mining operation in a dreadfully cold place.

In 1969, GCOS gave rise to yet another oil sands legend. A fleet of black limousines delivered a phalanx of GCOS executives to the legislature building in Edmonton to meet Manning's successor, Premier Harry Strom. Their message to Strom: we're losing money hand over fist, and if we don't get a royalty reduction, GCOS might not make it. Legislature wags suggested the Sun executives' message would've been more believable if they'd arrived in used Chevrolets.

In Alberta, GCOS's start-up agonies became common knowledge and contributed to a general sense that the case for oil sand development had yet to be made. Someone else would have to do the job. It was time for Syncrude to come on stage.

PART 2:
GROWING UP

4. THE LONG MARCH

It took a decade for the original Syncrude team to make the case for a major oil sand project and obtain regulatory approval from the Alberta government. It was a small and dedicated team—fewer than a hundred engineers, technologists, and dreamers—and the process involved years of pilot plant and laboratory tinkering, backroom drudgery, endless committee meetings with the owners, laborious consultations with government agencies and politicians, and not a few dashed hopes and disappointments before the final green light.

The Beginnings

At the end of the 1950s, three groups of oil companies were interested in building a full-scale commercial oil sands plant: GCOS, eventually owned by Sun Oil; Shell Canada, which contemplated an in situ plant to get oil from the deeply-buried Peace River oil sands; and a cobbled-together oil sands mining consortium called Cities Service Athabasca Inc., which was made up of Cities Service, Imperial Oil, Atlantic Richfield, and Royalite Oil.

Shell's up-and-down history in the oil sands was a barometer of European opinion. Remember that I earlier noted that in 1928, Shell attempted to exclusively lease half of Alberta and a good chunk of the Northwest Territories for oil exploration.

One of the original team: Ron Gray. (Photo courtesy of Syncrude.)

After the Oil and Gas Conservation Board deferred Shell's application in 1962, the company gained approval for a serious in situ project in Peace River and went on to make major investments in oil sands mining and upgrading projects from 1980 forward. In 2018, Shell's parent company in The Hague announced that it was beginning a major retreat from the oil sands as part of a worldwide rethink of its investments in petroleum.

Ron Gray, an irrepressible engineer from Regina, a man given to long, discursive sentences punctuated with sharp intakes of breath, was one of the very first people involved with the pilot projects and experimental enterprises that eventually became Syncrude. In 2011, he remembered the role played in Cities by Royalite Oil Company, a small independent producer that was bought out by British American Oil in 1962, after which it became Gulf Canada, an original participant in the creation of Syncrude in 1965.

Royalite begat British-American Oil, which begat Gulf Canada, one of the original venturers in Syncrude. If you were driving with the radio on in those days, you hummed along to this jingle:

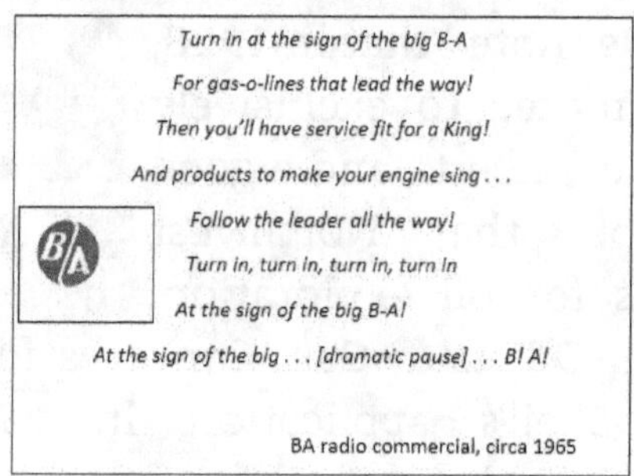

BA radio commercial, circa 1965

Recalling that Royalite's original experiments in the oil sands had included the work of an engineer named Gordon Paulson, who reputedly ruined his wife's washing machine trying to use it as a centrifuge to separate bitumen from oil sand, Gray said:

> So back in the mid-50s, we set up Lincoln Clarke, a very creative fellow . . . to do some research on the process using a centrifuge. The first time they put the diluted oil sand in this centrifuge . . . it looked like it was going to shake itself to pieces, [so] everybody evacuated the room. We then [switched] over to hot water extraction, and it seemed to work well. . . . Sometime in there,

all of us got carried away and made an announcement in the paper that Royalite was going to spend $50 million to build a 20,000-barrel-a-day oil sands project or something like that.[8]

Eventually, Gray said,

> We looked at a number of scenarios and came up with a big report examining all the details of the project and concluded that with crude oil at about $3.00 a barrel, in those days, the only way you could justify having a project would be to operate it at 100,000 barrels a day.

Around the end of June 1958, Charlie Hay, then Royalite's executive vice president, called staff into his office and announced, "We just entered into a deal with Cities Service to go ahead with the [oil sands] project." As Gray put it, "Well, Royalite didn't have the money, and I don't know what [we] were thinking of, frankly."

Gray's story is reminiscent of the legendary Saskatchewan junior hockey promoter Scotty Munro, who, it's said, organized a gathering of fellow promoters in a small-town hotel and concluded the meeting by saying, "All right, then, it's agreed. We'll sue the SOBs for a million bucks. Now how in hell are we going to pay for this room?"

Mildred Lake pilot plant test mine (Photo courtesy of Syncrude.)

Cities went on to do important research on hot water bitumen extraction and a new technology called hydrovisbreaking for refining bitumen into lighter petroleum fractions like gasoline, kerosene, and diesel fuel. Between 1959 and 1963, it operated a 1,000-barrel-a-day pilot plant at Mildred Lake, north of Fort McMurray, as well as a research laboratory on the east side of Edmonton.

At best, the plant could extract 1,000 barrels of bitumen a day. Initial recoveries were much lower. Mildred Lake was developed into a significant facility with Pan-abode cabins for the staff and a gravel airstrip that could accommodate the old DC-3 they used to fly in people and equipment. At its peak, the Mildred Lake pilot plant employed up to 125 people.

Building on its experience at Mildred Lake, Cities applied for approval of a full-scale oil sands plant producing 100,000 barrels of synthetic crude oil a day. At about the same time, GCOS applied for a 30,000-barrel-a-day plant and

Shell Canada applied to build a 100,000-barrel-a-day in situ plant in the deep Peace River oil sands.

Born in Disappointment

Cities Services Athabasca Inc., Mildred Lake pilot Plant. (Photo courtesy of Syncrude.)

In 1962, the Oil and Gas Conservation Board approved GCOS, the smallest of the three projects, and deferred the Cities and Shell applications for five years, sending both back to the drawing boards. Both applications, the board felt, threatened to swamp the market for Alberta's conventional oil production and force pro-rationing production cutbacks to conventional petroleum. The board —and a good part of the conventional oil industry—was opposed to that.

At the time, the United States had strict limits on oil imports, and Canada's national energy policy of the day confined western oil to markets west of the Quebec border. A sudden explosion of new oil production from the oil sands could conceivably swamp the marketplace and crowd out a significant part of Alberta's conventional production. GCOS won approval because it promised to limit its output.

In an alternative world, what would've happened had Cities been given the go-ahead instead of GCOS? The result might have been . . . unfortunate. Cities' pilot plant was less than a 100th the size of the 100,000-barrel-a-day plant they proposed. The consortium's mining expertise was minimal, its extraction know-how was incomplete, and its experimental hydrovisbreaking process for upgrading bitumen, though promising, had never been operated on anything larger than an 8-inch bench reactor at the Edmonton laboratory. Add to that the fact that the Cities project team of the day—fewer than 50 people—had never built anything even a 10th the size of the proposed complex, and it seems probable that a go-ahead would've resulted in huge problems during construction and operation. When a race is headed up a treacherous mountain trail, sometimes it's better not to be in first place.

The government's 1962 GCOS decision caused hardly a ripple in the oil industry itself. The oil world simply moved on. Every company needed new sources of oil to feed its network of refineries and service stations. The industry was like one of those huge whale sharks that sweeps up vast clouds of plankton and has to keep feeding, keep swimming, to stay alive. And so the industry went looking for new sources of conventional oil—in northern Alberta and northern British Columbia, in the McKenzie Valley and the Beaufort Sea, on the High Arctic islands, and, of course, in the Gulf of Mexico,

South America, Africa, and Russia.

Shell took its marbles and left to investigate other opportunities far away. Cities went into hunker-down mode. Most of the Mildred Lake staff were let go, and the pilot plant was closed. The last crew flew out on November 22, 1963. Indeed, they were on board the old DC-3 when they heard the news about the assassination of John F. Kennedy. The attitude of many industry leaders toward the oil sands was something along the lines of, "Sure, your time will come. Sooner or later. Come see us in a generation or two."

The four Cities owners (Imperial Oil, Gulf, Cities Service, and Atlantic-Richfield) decided to regroup and create a new joint venture called Syncrude Canada Ltd. In late December 1964, after an ice storm rendered Edmonton's streets impassible, Frank Spragins hopped, slip-slid, and ran the 12 miles from his suburban home to a downtown office building to be informed by the owners' representatives that he'd been named president of a new oil sands company.

Frankly, Spragins Gave a Damn

The young Syncrude was a very modest operation. More of an idea than an enterprise, it employed fewer than 50 engineers and technologists lodged in a run-down office above a downtown medical clinic and several second-hand ATCO trailers on the edge of Edmonton's refinery row. All kept alive on a small financial lifeline from the company's four participants.

Did Spragins think he'd just been made captain of a sinking ship? Almost certainly not. He was very much a glass-half-full kind of man, very attuned to opportunity.

Following his appointment. he'd devote the next seven years to cementing the coalition, developing a plan for a more advanced oil sands plant, and making the case that the Alberta government should change oil sands policy to allow more rapid development. If he hadn't done this work and done it so well, Syncrude might well have become a historical footnote. As it was, Spragins helped guide the project to its completion and start-up before he died of a brain tumour in 1978, scarcely two months after its official opening. He worked until two days before his death.

Some of Spragins's advocacy was directed at the big kahuna of Alberta regulatory bodies: the Oil and Gas Conservation Board. Some of it took the form of newspaper interviews and speeches to business and conference audiences. Much of it was years of tedious, endless committee meetings in Toronto, Tulsa, New York, and Los Angeles. All of it, you might say, was endless rounds of unspectacular gut work.

Spragins was a soft-spoken and considered man who thought and read before he spoke. This served him well in the committee and boardroom. Privately, he was an avid reader, gardener, hunter, and angler. He was trim, physically active, and prematurely bald in his mid-thirties, which always made him seem older than his years. He was something of a polymath.

Clockwise from top left: Spragins in his twenties; at work as Seismic crew chief; the angle and camper; in the early Syncrude lab.

He was born Franklin Keller Spragins in Natchez, Mississippi, in 1914. His father died in an industrial accident when he was only a baby. The family wasn't desperately poor, but they counted their pennies. His mother remarried, and Spragins didn't get along with his stepfather; he was therefore sent away to be raised by two different aunts. The second lived in East Texas, which is where Spragins attended high school.

He got a scholarship to study engineering at Rice Institute (later Rice University) in Houston. He later recalled that Rice was a tough place to study; in most courses, the passing grade was 95 percent. He graduated in electrical engineering in the late 1930s and got a job with Carter Oil, an Exxon subsidiary. They sent him to Canada in 1942 as part of the US war effort to find new sources of oil in northwestern Canada (because of the Japanese threat).

Spragins wasn't prepared for Canadian winters, recalling to me that during his first winter in southern Alberta, when the temperature was 30 below, he walked to work in shirtsleeves and froze the top of his head (whether this contributed to his early baldness isn't known).

He spent a number of years working as a seismic crew chief for Carter and then, eventually, for Imperial Oil in Calgary. He did exploration work around southern Alberta as party chief of a seismic crew and narrowly missed participating in Imperial's famous Leduc #1 discovery that kicked off the Alberta oil boom (he was working on another project nearby). In 1949, Spragins joined Imperial's department for oil sands initiatives and became its first manager. In this capacity, he participated in Imperial's early ventures, including its association with Cities.

Spragins a few years before his death. (Photo courtesy of the Spragins family.)

He was with Imperial Oil's exploration department in 1949 when he met Eleanor (Nell) McGregor, a pretty stenographer in the department. McGregor, who described herself as a Turner Valley farm girl, liked the "really nice" American oil explorationists she met there and was attracted by Spragins's soft voice and gentle accent. They married the next year and remained so until he died.

Spragins wasn't the first or only man to have "fallen" for the allure of the oil sands; don't forget Sidney Ells, Robert Cosmas Fitzsimmons, and, of course, J. Howard Pew. What set Spragins apart from all of them was his vision of all the economic and social development the oil sands could support. He had a deep appreciation for the urgency of developing the sands before conventional oil ran out and other sources, like the Colorado oil shale, were developed. And perhaps more important, he believed that the sands should be developed in a way that broke the poverty barrier that prevented Indigenous peoples from participating in the industry's full benefits. Spragins was the first to suggest that the oil sands industry should be socially responsible. In his mind, *how* they should be developed was as important as *whether*.

Holding It All together

For a number of years, Spragins provided much of the glue that held the sometimes-shaky Syncrude coalition together. The participants, as they called themselves, blew hot and cold on the oil sands and were influenced greatly by industry developments in other parts of the world. A major example was the 1968 discovery of a so-called elephant oil field at Prudhoe Bay, Alaska, which sent a shock wave through the industry, especially affecting two participants in Syncrude: Esso and Atlantic-Richfield. It took nearly six years to play out, but the consequences of Prudhoe Bay eventually led ARCO to pull out of Syncrude and almost capsize the coalition.

Spragins's role with the participants was a little like Dwight Eisenhower's as supreme allied commander during the war, namely holding together a coalition of independent people with their own goals, cultures, egos, hopes, and fears. There was a reason Syncrude had four participants: none of the companies was prepared to take on the project's risks by themselves.

At different times, Spragins told me, every one of the four participants had been ready to throw in the towel until he told whoever it happened to be that at least one *other* participant "was getting really hot.... If I heard ARCO was getting nervous and thinking about pulling out, I'd tell them, 'Gosh, Cities is getting really excited about the project,' and that would bring them back because none of them wanted the others to get in on a good thing that they had passed on."

How Syncrude's "Strange Structure" Made Great Things Possible

Syncrude proved to be a very creative enterprise because its peculiar corporate structure created an opening for the liberation of people's creative energies. Leaders who knew how to take advantage of those openings turned Syncrude into a very innovative place, a leader in megaproject management, workplace innovation, a truly innovative employee relations culture (the country's first), and in forming partnerships with Indigenous peoples, which we'll explore in greater depth as we continue.

Think for a moment of companies thought to be innovative, creative, and transformative. You're probably imagining the likes of Amazon, Google, Apple, Facebook, and Netflix. Or maybe you're thinking of companies from a couple of decades earlier, like Northern Telecom or Blackberry. You probably didn't picture a major oil company. Why was Syncrude destined to be different?

Part of the answer was leadership, and another part good hiring. Still another part was a corporate structure that was almost a "clean sheet of paper." Syncrude wasn't shaped by the culture of any of its four owners. It was able to invent its own.

On paper, the company was managed and directed by an executive committee of officials from the four owners; their votes were proportional to their ownership shares (i.e., Imperial Oil held 25 percent of shares and 25 percent of votes). In practice, Syncrude management had an enormous amount of autonomy. Once the construction go-ahead was announced, the sheer size of the company's tasks and the urgency of getting the project built and operating (in less than five years) meant that its people had huge latitude in making myriad decisions without the owners' involvement. And as future CEO Brent Scott liked to say, it was always better to ask for forgiveness than permission.

On paper, Syncrude's executives took direction from the owners in monthly executive committee meetings. In reality, once the Syncrude ship left the harbour every month, the owners could certainly radio the captain and his bridge officers, but most of the time, the captain and crew ran the ship themselves. The owners saw Syncrude as their creation, and speaking as someone who once worked for Syncrude, I can say that most of us truly believed that Syncrude was *our* company—and we acted like it.

I don't mean to suggest that the owners

didn't contribute to the company's success or that the contributions of their representatives were unimportant. Most of Syncrude's senior executives—Spragins, Projects SVP Chuck Collyer, Operations SVP Neil Lund, and first HR Director Don Scott—were all Imperial Oil people. Herb Shepard, the academic and industrial psychologist who became the guru of Syncrude's team management culture, came to the company via Imperial and Exxon. The owners contributed other senior personnel and paid the bills in the long years before Syncrude generated income, and staggering bills they were. Syncrude was the biggest megaproject in the Canadian fossil fuels industry for an entire decade. At the peak of construction, it was spending a million dollars a day, and in those days, to coin a phrase, a million here and a million there pretty soon added up to real money—$8.2 billion in 2020 dollars to be precise.

It's certainly true that Syncrude's management and staff couldn't have accomplished what they did without the support of the many owner representatives—especially Floyd Aaring of Gulf and Richard Galbreath of Cities—who made timely and responsible decisions and effectively kept their own senior managements onside and supportive.

How Syncrude's people put that freedom to work, I'll deal with shortly.

Selling the Dream

Spragins's second contribution was his imaginative and persistent selling of the possibilities of the oil sands to industry, government, and the public.

It's hard to imagine today, but prior to the 1980s, the majority of major oil companies were deeply skeptical of the oil sands proposition. Many industry people didn't view sands types as *real* oilmen. Indeed, in the early 1970s, a senior executive of Sun Oil was nearly thrown out of the Calgary Petroleum Club after a member complained that he *wasn't* a real oilman. Of course, the Calgary Petroleum Club wasn't exactly a model of cultural enlightenment: women weren't allowed in before 5 p.m. so "oilmen" could meet for lunch and hammer out deals without impediment.

People in the conventional oil industry —companies that searched for underground deposits or made their living drilling them and refining the oil into marketable products like gasoline, diesel, and aviation fuel—were skeptical that investing in the oil sands would ever make business sense. For a long time, it seemed they were right. GCOS's struggles were proof of that. At the time Syncrude began construction in 1972, the price of Western Canadian light crude oil was less than $3 a barrel, and few people thought oil sands petroleum could be produced for anything close to that.

A large part of the oil industry was deeply wedded to searching for conventional oil reservoirs (the kind you tap by drilling wells). The industry itself—not to mention the public—needed reasons why the sands should be developed *now* rather than some indefinite time in the future. That was why, against all the negative news of the 1960s and early 1970s about GCOS, Spragins's missionary efforts to sell the needs and benefits of oil sands development was crucial.

One of the retail tools Spragins developed—which became semi-legendary in the Alberta of those days—was a 15-foot-wide by-products chart that painstakingly displayed dozens of substances, everything from silica sand to sulphur, radioactive isotopes, and heavy metals, all of which were contained in and could potentially be recovered from oil sands. By-products, he argued, could turn northern Alberta into a second Ruhr. Remembered Ron Gray:

> Frank would say, 'Look at all the advantages this could bring to Alberta if we had more, not just the ethylene, the propylene and the butylene, but there is sand from which you can make glass and all sorts of rhenium, zirconium, and titanium associated with these clays . . . when you float the froth, you

float up all these other minerals.

To the best of my knowledge, the only by-product that the oil sands industry has developed commercially is elemental sulphur, which is today used as the raw material for making ammonium sulphate fertilizer. It's possible that in the more distant future, some part of the oil sands will be increasingly used for the manufacture of carbon products like carbon fibre, activated carbon, graphene, carbon nanotubes, metal carbides, synthetic graphene, or hydrogen. Should that happen, one hopes that Spragins's early work will be remembered.

A Leg Up for Indigenous Peoples

Perhaps Spragins's largest contribution was his advocacy for using the oil sands to finally end the poverty cycle that had locked so many Indigenous peoples outside the modern economy. He was the first to see that the oil sands presented an opportunity to advance social change as well as economic development. He was also the first oil executive to be seriously interested in Indigenous peoples and in the need to shape oil sands policies that would create job and training opportunities for them. To that end, he was the first to put effective industry programs in place.

Spragins had a very sociological understanding of the challenges his vision would face. He understood that most Indigenous peoples in the northeast had little experience of the industrial economy and didn't understand its requirement not just that an employee come equipped with an industrial skill set, but that they show up for work consistently and on the employer's schedule. At the time, the perception of many white business people was that Indigenous peoples weren't reliable employees, and shifting that perception was going to require change on the part of both employers and employees.

Spragins was acutely aware that Alberta's past economic development had largely ignored Indigenous peoples, and he was determined that, when Syncrude's turn came, things would be different. During the 1960s, when Syncrude was only one of several prospective oil sands developers—and not the first to get a go-ahead —Spragins reached out to Indigenous leaders across northern Alberta and worked with them to encourage the province's first pilot projects to train and employ their people. He got involved in and helped promote Alberta Newstart, an innovative adult-education project based in Fort McMurray. He knew, liked, and was trusted by many of the community-level Indigenous leaders in northern Alberta, a fact that played a significant role in Syncrude's initiatives in that area.

Spragins was an effective negotiator and dealmaker. A rarity among engineers of his day, he was interested in the world of politics and took time to study and get to know politicians. As far as getting government approval for his oil sands project, he was determined to get the politics right. How and why he had this realization and went about putting it into effect is something we'll deal with in a later chapter.

The Comeback

Any account of Spragins's treatment by historians must address the dishonest and unbalanced portrayal of the man in an infamous 1976 CBC docudrama entitled *The Tar Sands*. This program, which was aggressively promoted by the national network, was a low point in its coverage of energy issues at the time.

Peter Herrndorf, an ambitious young CBC producer, commissioned the writing of the special, which portrayed other crisis-mode negotiations between Syncrude's owners and Alberta Premier Peter Lougheed in 1973 and again during the ARCO pullout in 1974.

The oil company representatives were portrayed, mustaches and all, as sleazy suits. The character of Spragins was portrayed by an old Stratford theatre actor and director, Mavor Moore, as a fat, back-slapping, four-flushing "Texas oilman." This was a deep injustice to a decent man who was none of those things. Spragins was a quiet, almost courtly man, an industry statesman who was decent, honest, and unfailingly diplomatic.

I watched *The Tar Sands* with Spragins, his family, and Scott, then the company's new CEO. I told Spragins he should sue the CBC for libel, especially since he was portrayed by name (the owner representatives were all anonymous) and wasn't a public figure as the courts then defined one (the courts took a dim view of anyone libelling a private citizen).

"You don't put out a fire by pouring gas on it," Spragins said quietly. He believed that as a CEO, he had a responsibility to keep his ego out of public debates.

Lougheed felt differently. *The Tar Sands* portrayed him—not unfairly, on the whole—as a provincial leader caught up in a high-stakes negotiation, the first of the energy era, doing his best to work through a complex situation. Some people who knew Lougheed well thought Ken Welch, the Alberta-born actor who played him, portrayed him fairly enough as a man grappling with uncertainty, pushed back and forth between his advisors on the one hand and the aggressive oil company representatives on the other. A man determined to get fair value for the oil sands.

Lougheed disagreed. Whether his ego was offended or he thought he had to show Albertans that he wasn't going to take any guff from the national broadcaster, Lougheed filed a lawsuit against the CBC and brought the resources of the Alberta government to bear on prosecuting it. The CBC eventually surrendered, giving Lougheed a cash settlement and a promise to bury the show in its vault and never air it again. Hopefully, the premier donated the cash settlement to the public treasury.

5. A FUNNY THING HAPPENED ON THE ROAD TO FORT MCMURRAY

For Syncrude's people, to say nothing of its owners, the five-year construction period was a time of very nervous stomachs. The world of oil had lost its stability: the 1973 Arab-Israeli war had led to a cut in Arab oil exports to North America and skyrocketing oil prices, which in turn induced massive inflation. And then, without warning, Atlantic-Richfield, one of the four original Syncrude participants, dropped a bomb on the project.

In early December 1974, Sam Stewart, ARCO's representative on the management committee, walked into the committee and made an announcement that shook the meeting. Stewart was white-faced, as well he must've been given that what he was about to announce would quite possibly doom the Syncrude project and likely spell the end of his own career. Looking back on the meeting, Scott said:

> Visualize a meeting of the management committee, and they're all sitting there feeling, you know, they're talking and joking and having a reasonable time, and Sam Stewart . . . he was a particularly agreeable kind of a guy. He was very calm and cooperative and so on. And he came to the door on this morning, and he really looked horrible, and he said, "We're pulling out." He had got the bad news from head office, I guess. They were leaving $18 million dollars on the table, so to speak, and they were going to leave, and that was because of problems or requirements from the [Alaska and] North Sea oil production, which they were big in. He looked absolutely devastated, and he left and the rest of the group looked devastated too.

ARCO was in a difficult position. It had an enormous stake in the Prudhoe Bay oil field in Alaska and a major pipeline to bring the oil to market. Costs were escalating. The company needed to rationalize its investments. Shockingly, the original joint venture agreement among the four participants had no prenuptial agreement; for some reason, no one thought to negotiate one. ARCO could walk without penalties, other than leaving behind the cash already invested.

Looking back, perhaps ARCO's decision shouldn't have come as a total surprise to Syncrude management and the other three companies; after all, they'd just been through harrowing negotiations with the Alberta government to try to establish an economically attractive royalty scheme. Besides that, even though construction had just begun the cost estimates for the project were already escalating, and the oil industry was worried about the consequences of the Arab oil embargo. Plus, there must've been murmurs coming out of ARCO in the months earlier about how the company was going to cope with the alarming costs of its planned Alaska pipeline.

Even so, the decision was a thunderbolt. In a single day, Syncrude went from the hottest business story in the country to a likely candidate for delay or even shutdown and cancellation.

ARCO's pullout rocked Syncrude to its foundations. It threatened to not just stop the construction project and leave it as a skeletal hulk, a monument to folly, but to also trigger a raft of billion-dollar lawsuits and set back the cause of oil sand development by a decade or a generation, if not forever.

Syncrude was looking over a precipice. New employees bailed out or started looking for new jobs. After all, who wanted to board a sinking ship? Job offer letters were put on hold. Stresses soared. While the owners' debated what to do, anxious employees agonized over their bills. Should they hunker down and hope for the best or ..? Some tried to calm their families and scan the want ads at the same time. Others drank.

ARCO was responsible for 30 percent of Syncrude's costs. The three remaining oil companies—Imperial, Gulf, and Cities—couldn't be expected to simply ante up the millions that ARCO was walking away from. Their businesses were already exposed to as much oil sands risk as they could tolerate. Thus ensued 60 days of desperate behind-the-scenes attempts by the three companies to try and find an additional investor to pick up ARCO's tab.

Rescue in Winnipeg

In late January, Bill Mooney, a Cities VP, flew quietly into Winnipeg, booked 30 rooms at the International Inn hotel, and then flew home again. It was a curious thing, unnoticed at the time. Why Winnipeg? Well, why not? If there were to be talks on the future of Syncrude, let alone the oil sands, best they take place on neutral ground, not in Alberta or Eastern Canada. Winnipeg was neutral ground, kind of like hosting East-West disarmament talks in Iceland.

Mooney, the so-called shuttle diplomat. (Photo credit: The Mooney family.)

Mooney—a school athlete who'd graduated from Father Athol Murray's famous private academy in Notre Dame, Saskatchewan—knew a thing or two about survival in tough times. Notre Dame was famous for operating on a shoestring and producing ferociously competitive sports teams.

The negotiations that led to the Winnipeg Agreement, which encapsulated this understanding, have been described more fully elsewhere,[9] but here's the gist. By 9:30 a.m. on February 3, the delegations were in place. They included the federal Treasury Board's Jean Chretien and Energy Minister Donald Macdonald, Premier Lougheed, and Premier Bill Davis of Ontario.

Chairs were reserved for Shell Canada's President Bill Daniel and his executives, but within a short time, Shell announced it would only consider investing in Syncrude if government would guarantee a minimum floor price for synthetic crude oil. This was unacceptable to all parties, so Shell quit the talks. That left three companies and three governments at the table.

Mooney, at the time a little-known player with Cities, became the game's unsung hero. A garrulous relationship-builder, he played a vital shuttle diplomacy role behind the scenes, facilitating communication between the oil companies and especially Minister MacDonald. Mooney deserves great credit for lubricating the deal-making process between oil company execs and government officials who had little trust in one another.

As Mooney's obituary put it, "In all negotiations and cooperative efforts throughout his career, Bill held true to one over-riding belief —that if he put the interests of his country first, his province second, and his company third then every solution would be beneficial for all parties involved."

Over a period of days and weeks, a deal gradually came together. Mooney kept on with his behind-the-scenes shuttle diplomacy, getting people together, cajoling them into agreement, and searching for common ground. Issues that were worked out included the price of synthetic crude, Syncrude's contracts with Ontario suppliers, and, for Canada, energy self-sufficiency. In all likelihood, the talks encouraged a growing Ontarian and federal awareness of Syncrude's potential to create jobs in Eastern Canada. The talks reinforced support for Alberta's 50-50 net profit royalty system, and the province agreed to take a major ownership position in the synthetic crude pipeline from Fort McMurray to Edmonton as well as in the utilities plant to support the complex, both extremely lucrative and low-risk investments.

Over the years, several green or leftist writers have alleged that the Winnipeg Agreement amounted to American oil barons outfoxing Canadian politicians in order to suck money out of Canadian taxpayers. This belief flew in the face of the widespread benefits to Canadians that flowed from the agreement, which were far in excess of what it cost government.

By any calculation of the pluses and minuses (direct and indirect jobs, government revenues and ancillary benefits), the Winnipeg Agreement was a splendid demonstration of Canadians' basic pragmatism and willingness to invent private-public partnerships to underwrite nationally beneficial works with broad social benefits. In the end, the feds took up half of ARCO's 30 percent share of the project. Alberta took up 10 percent and Ontario 5 percent. All three governments eventually parlayed their investment into major profits.

Syncrude dodged a bullet.

In Edmonton and Fort McMurray, hastily written shutdown and abandonment plans were filed away, and people happily plunged back into their original tasks. And the Syncrude-Bechtel team went back to the jobs for which they'd been hired: to keep the project moving steadily ahead on the rails.

Of course, their challenges had just begun.

6. THE BIG, TOUGH, EXPENSIVE JOB

When Harold Macmillan became British
Prime Minister a reporter asked whether
he was worried about anything.
"Developments, dear boy!"
he said. "Developments!"

In less than a decade, Syncrude grew from a tiny research and development company operating on the fringe of the oil industry into a major Canadian enterprise with more than 5,000 employees and the skills and resources needed to build and bring into operation the world's largest unconventional oil production project.

As we've seen, the early parts of that story took place under Frank Spragins's leadership. The later parts and their culmination took place under Syncrude's second president, Brent Scott. In 1974, recognizing that Spragins had no experience with building a major project or shaping a major enterprise, the owners elevated Scott, then senior VP, to the role of CEO, and they moved Spragins to the position of non-executive chairman. It fell to Scott to lead the creation of the strong human organization that could build and operate the complex, and it was also his responsibility to keep the owners on track from 1973 until launch in 1978.

Syncrude was destined to be the biggest oil industry megaproject in the first 70 years of the 20th century. Everything about it ventured into unexplored territory.

The plant 40 kilometres north of Fort McMurray would be three times the size of GCOS. To feed its appetite for oil sand, it'd dig the largest open-pit mine in the world. To extract the bitumen, it'd need the largest mineral-processing facility in the world. To upgrade the bitumen into synthetic crude oil, it'd use two of the biggest oil refining units (called "fluid cokers") in the world. To generate the steam and electricity needed to power all this processing, it'd need a natural gas–fired power plant big enough to light a city of 300,000 people plus a high-voltage tie line to the Alberta power grid, which in turn was connected to the Alberta, British Columbia, and West Coast energy grids.

To bring one of the biggest single public works of the 20th century into existence would require the skills and energy of up to 8,000 on-site construction workers representing 21 skilled trades, thousands of miles of steel pipe, hundreds of thousands of miles of electrical wiring, and high pressure steel vessels from France and Italy. Along with hydrogen furnaces from England, bucket-wheel mining machines from Germany, four of the largest dragline mining machines in the world from two plants in Ohio and Illinois, centrifuges from Sweden, and cranes, pumps, valves, nuts, bolts, and about 10,000 other components from a dozen countries, not to mention a global network of ships, trains, trucks, aircraft, assembly yards, and warehouses to store and ship all this stuff to the Syncrude site in the northeastern Alberta frontier, all of which had to be done on time and on budget.

The job of organizing a staff and contractors to accomplish all of this—to get it designed, contracted, gathered, and finally assembled—fell to Syncrude's Projects Department and to the company's new senior vice president of projects, C.R. (Chuck) Collyer.

Like a successful army, a major enterprise needs leaders with vision, staff officers who can translate that vision into detailed workable plans, and intelligent, dedicated combat soldiers who can carry out the plans on the ground and against resistance. Syncrude was blessed with all three.

While this account focuses on the leaders, one must always remember that Syncrude's

Collyer in 1975. (Photo courtesy of Syncrude.)

successes were the product of extraordinary teamwork and the dedicated efforts of thousands of people—engineers, technologists, skilled tradespeople, planners, logisticians, social scientists, environmental officers, and myriad other knowledge workers.

The profound importance of team thinking in what Syncrude was able to accomplish was, in my view, one of the company's most valuable legacies. At Syncrude, teamwork wasn't a marketing slogan—it was a deeply thought-out corporate culture that reached from the president's desk to every corner of the organization. It was a way of approaching work every day; you might say it was a way of life.

A Man for the Season

Chuck "only my mother called me Charles" Collyer grew up in Montreal, graduated from Queen's in engineering in 1952, and joined Imperial Oil. He was every inch an Imperial man, having spent his career first working in the Montreal East refinery and then moving to manage the construction of company projects in Sarnia, Ontario, and Alberta.

In those days Imperial Oil was the largest Canadian oil company and the largest of Syncrude's four owners. It had the industry's biggest talent pool, or what football coaches would call "the deepest bench." In the early years, it was therefore understood by the four owners that Imperial would contribute the largest segment of senior managers to Syncrude. Two of Syncrude's original leadership team, Collyer and Neil Lund, were Imperial people; so was Syncrude's first human resources manager, Don Scott. President Brent Scott, of no relation, came from Gulf Canada.

Collyer was highly regarded at Imperial. In the late 1960s, he was a senior project manager at Imperial's Judy Creek gas plant and then the company's big Redwater fertilizer plant. In 1971, Imperial picked him to work on the early tasks of preliminary design and cost estimating of the Syncrude project. Collyer was called to Toronto to meet with the president of Imperial, R.G. "Dick" Reid, and he was offered the job as senior project engineer. He was experienced enough to have some sense of the size of his task, and he was a can-do kind of guy. Naturally, he accepted.

> I was 42. My initial assignment was
> in 12 months to firm up the scope of
> the project and develop a cost estimate

> and schedule for the owners to consider their commitment to proceed. . . . When I asked the current outlook for the project, oil was $3 a barrel. Dick said, based on some preliminary studies and geological data, that Syncrude's cost was estimated at $350 million. If it ever got to $750 million, he told me, it was "ball game over."

Reid's forecast proved deeply ironic. At the end of Collyer's 12-month study, Syncrude's revised cost estimate had grown from $800 million to $1.2 billion. A year later, when Syncrude was into construction, a further revised estimate placed the total cost at $2.1 billion. Despite these rising forecasts, the project rolled ahead. None of them led to the ball game being called (although the ARCO crisis came close).

By several orders of magnitude, the job would be the biggest assignment of Collyer's life. Lund had managed the Sarnia refinery, which had a workforce in the hundreds; he'd be charged with building an operations team of more than 4,000. The same stretch was true for many others, including me.[10] None of us knew exactly what we were getting into.

Said Collyer,

> Only after [the Winnipeg Agreement] . . . were we given the release to resume engineering and

procurement. This required a change in execution strategy, and only an incredible performance by the project team enabled completion to occur on schedule and essentially on budget—something not achieved on subsequent oil sands projects. We were left with the hesitancy of the market to buy in to the "go" reality. For example, contractors were hesitant to commit to building infrastructure in Fort McMurray, and prospective employees were reluctant to relocate to Fort McMurray.

In those days, military officers and business executives knew the importance of physical presence. Certainly, Collyer did. He was a big man, 6'1 or so, and imposing. He didn't swagger exactly, but he conveyed an air of smiling, unassailable confidence just a tad short of cockiness. He'd usually arrive a few minutes late for meetings and make a dramatic entrance, going up to his chair and making eye contact with everyone around the table before sitting down slowly and settling, as if to say, "Okay, I'm here. You can start now."

One of the first priorities of the Syncrude executive (and of course the owners) was to hire a world-class managing contractor: a company global enough, experienced and competent enough to coordinate the thousands of contractors, subcontractors, and suppliers and tens of thousands of workers that would be needed to build the Syncrude complex. In the 1970s, there were only three such managing contractors in North America, all of them based in California and all of them *very* American. The acknowledged industry leader was Bechtel Corporation of San Francisco, California.

No one questioned Bechtel's competence or its track record. It built the Hoover Dam—the signature public work of Franklin D. Roosevelt's New Deal in the Depression—and in the 40 years following, it built an astonishing array of the world's major oil refineries, petrochemical plants, pipelines, hydroelectric dams, nuclear power plants, urban rapid transit systems, ports and airports in dozens of countries around the world—not to mention GCOS itself.

Bechtel was the 900-pound gorilla of the construction and engineering business. Its people were good at what they did, and they knew it.

A Crucial Decision

While Bechtel may have had the inside track on account of its size and performance record, in Alberta's engineering and business community its management was seen as arrogant and deaf to local sensitivities (especially the sensitivities of Albertan engineers).

There were many people who remembered

how Bechtel's US managers had made little provision for Alberta or Canadian engineering firms in building projects in Alberta in the 1950s and 1960s, starting with GCOS. A minor legend in Alberta's business community was that in in the course of building GCOS, Bechtel had imported a gravel crusher from California. A *gravel crusher*! Because, obviously, you couldn't expect the rubes in Alberta to know how to find a gravel crusher locally.

Many years later, leftist journalist Sally Denton attacked Bechtel's lobbying prowess and deep links to right-wing politicians in the United States and abroad, describing the company as a manipulative Cold War "deep state" closely linked to the Republican right and the CIA.[11] But in Alberta in 1972, that critique was years away What people talked about were stories of Bechtel's conduct in the province during the 60s. Spragins said to me, "[Premier] Manning told me, 'Whatever else you do, don't hire Bechtel.'" He smiled ruefully. "So of course, the first thing we did was hire Bechtel."

Brent Scott recalled how the managing contractor selection process began:

> The owners had sent out a request to four [of the world's leading managing contractors] for proposals. [For reasons I have never understood,] the proposals shipped to a hotel in Denver. And so the

> management committee, we all went down to Denver, and there were the proposals ... they were [motions a stack a foot high] that high. We had a little meeting and decided, well, the only way to handle these damn things is to get them up to Edmonton and let our guys get at them, so we quickly arranged to fly back to Edmonton. Each of us carried about 40 pounds of proposals onto the plane.[12]

In Collyer's mind, and probably Brent Scott's and the owners' too, Bechtel was neither the enemy nor a necessary evil: it was a potential vital partner without which an enormous and complicated task couldn't possibly be completed. Certainly, Syncrude couldn't hope to design and build the job itself: at the time, it had fewer than 150 employees and zero corporate experience managing big projects. Getting the right managing contractor was a way to reduce the enormous risks of a megaproject. Plus, as they used to say, nobody ever got fired for recommending IBM.

Looking back on the selection decision in 2020, Collyer recalled:

> The selection of Bechtel was not an arbitrary decision but one in which I had involvement in 1971, beginning with contractor screening. My first

involvement with Bechtel [went] back to the Montreal IOL expansion in the 50s, Sarnia in the early 60s, and Redwater in the late 60s. I had the benefit of knowing people at all levels of the organization and evaluating their performance and their management systems. This created mutual respect and trust. It allowed me to reject [Bechtel's first recommended project manager, who had project managed GCOS] until I got who I wanted.

[The Bechtel man I did want] and I had worked very closely and successfully on Redwater, and I was confident we could structure a closely integrated project team which would be unique but essential to successfully executing the project. Bechtel had the resources, experience, and management systems critical to the needs of the project but offered the potential to create the integrated client/contractor relationship I deemed necessary for a megaproject. The success "we" created is very much due to the outstanding collaboration by Bechtel and its people.

So Bechtel was chosen. If Collyer was at that time unaware of Bechtel's reputational baggage, he certainly became painfully aware during a dinner sponsored by the Association of Professional Engineers, Geologists, and Geoscientists of Alberta (APEGGA). "The project environment was hostile," he said. "At a dinner meeting hosted by APEGGA to explain our Canadian content strategy, not a single attendee had the courtesy to talk to me throughout [the entire evening] dinner." Chuck Collyer was a proud man. Forty-seven years later, that still stung.

Spragins and Brent Scott took the brunt of the selling job to the engineering community. Explanations were offered and ideas solicited. Meetings were held with Canadian engineering and construction firms. Proposals for increased Canadian and Albertan content were tabled and discussed. Over the months that followed, Bechtel somewhat Canadianized its Syncrude team, major Canadian and Albertan engineering organizations were brought into the work, and major contracts were signed.

In 2020, Collyer looked back on it all:

> In principle, we were committed to using Canadian companies where they had the necessary experience and resources . . . [so we hired] Monenco for the utility plant, Poole Construction for heavy civil engineering and associated engineering, supplemented by SCL and Bechtel technical personnel, for the

extraction plant. These were significant assignments and also augmented by hundreds of other subcontracts. . . . I believe an exceptional job was done in maximizing Canadian content at every level.

How this maximization was done, exactly —how the details of it were hammered out between Syncrude and Bechtel and between Bechtel and the Canadian contractors and how the Canadian content of the project was measured—is lost to time. Suffice to say, eventually the Canadian and Albertan engineering communities came to be satisfied that Syncrude had made its best efforts and that opportunity maximization was done reasonably well. And the experience the Alberta and Canadian industry gained left a strong legacy of involvement in future oil sands projects, of which there were a multitude.

A different issue, on which Syncrude and Bechtel representatives initially collided, was jobs for Indigenous peoples during construction. This issue came to a head in 1974 when Syncrude's community relations coordinator, Terry Garvin, had a very bad meeting with Bechtel's first on-site construction manager, a man who, since he's not here to defend himself, we'll call "Mister C."

When Garvin told him of Syncrude's goal for

Indigenous hiring, Mr. C growled, "They can *have* a job—mine." He threw Garvin out the door. This incident precipitated a project review meeting in San Francisco between Syncrude executives and Bechtel's senior management, after which Mr. C was replaced. And Bechtel resolved to support Syncrude's goals to hire Indigenous peoples.

This was the first of several points at which Syncrude wrestled with how to measure Canadian content, Albertan content, and Indigenous content. Bechtel employed many Canadian engineers and other workers. Were they included in the count? The same problem of statistical counting arose in Syncrude's Indigenous hiring (or as it was called in those days, its "Native" hiring). This was, after all, a time when it was illegal to ask the ethnic origin of an employee, and when there already existed in law such categories as "status Indian," "non-status Indian," and "Metis." How were we to go about determining how many Indigenous employees were hired? In that case, we simply invented a reasonable category called "Native employee" and applied it to people with surnames known to be widespread among people with some degree of Indigenous ancestry. A small sample of popular surnames included Ducharme, Boucher, Cardinal, Desjarlais, Ladoceur, and Ermineskin.

If Collyer was intimidated by the scope of the challenges ahead of him, he didn't show it. His team and Bechtel had to get their arms around developing a detailed design for the project, negotiating hundreds of contracts for the building of all the pieces, pulling them together from around the world and onto the project site north of Fort McMurray, and then connecting hundreds of complicated machines, controls, and instruments and the tens of thousands of components, pipes, plumbing, and wiring into a complex capable of mining vast quantities of oil sand, extracting the bitumen, upgrading it, and, finally, pipelining 125,000 barrels a day of synthetic crude oil to markets across North America.

They had less than five years to do it. In a world that was starting to go haywire.

* * *

Months earlier, on October 6, 1973—Yom Kippur—Egyptian and Syrian forces launched an attack on Israel that sparked a 19-day war that almost led to a superpower nuclear confrontation. What it did lead to was revolutionary changes in the world of oil, changes that still reverberate to this day.

At the end of the war, the oil-producing Middle Eastern countries launched an export embargo against the United States, Canada, and several European nations. This embargo triggered runaway inflation and a multi-year energy crisis in the West that washed across its economies and shook the North American oil industry. It also triggered a gasoline supply shortage in the United States. By February 1974, 20 percent of US gas stations had no fuel. Lineups and ugly incidents at gas stations made the news. Rationing was imposed. Truckers were shot at and even bombed.

By the end of the embargo in March 1974, the price of oil had risen nearly 300 percent, from US $3 a barrel to nearly US $12 (US $61 a barrel in 2018 dollars). The embargo caused an oil shock, with many short- and long-term effects on global politics and the global economy. In the United States, a national maximum speed limit of 55 mph was imposed. A Strategic Petroleum Reserve was created, and so was a federal cabinet–level Department of Energy. There were calls to conserve energy and then to rapidly develop alternative energy sources, including nuclear, renewable, and domestic fossil fuels.

Suddenly, the issue of North American energy self-sufficiency, which had hitherto mainly concerned policy wonks, was everywhere. It provided oil sand development with a tailwind. But there was a headwind too: the soaring prices of everything needed to build an oil sands plant, including steel, rubber, electricity, and every kind of skilled labour.

Oh yes, one other piece of context for the history of Syncrude. It was the 1970s, which meant the team had to design and build the complex without computers, digital communication, or the internet. Older engineers still used slide rules; younger ones used electronic calculators. Syncrude's chief (and only) economist used a hand-crank calculator. (Even in those days he was considered a bit behind the times, an impression that he reinforced with his penchant for hand-writing reports in an antique script).

And so Collyer, the Projects Department, and Bechtel embarked on a four-year quest to build the big, tough, and expensive job of their dreams. And they wrote some history while doing it.

First Steps

In stiff, project-management language, Collyer recounted:

> With time a significant cost factor, timely decisions and approvals would significantly improve the engineering and procurement process. For that reason, I wanted experienced Syncrude personnel resident in key offices with the appropriate authorities to avoid unnecessary recycle and expedite their review and approval function. Most [of these Syncrude people] had performed this role before. The design teams across the project would be working

to Syncrude-supplied mechanical design specifications. To facilitate and provide greater design basis clarity I had a "design basis memorandum" prepared with guidelines on important issues such as provisions for future expansion, fire and safety, etc. This was jointly prepared by Syncrude and Bechtel in anticipation of design issues commonly encountered in the engineering phase.

And so the Syncrude-Bechtel team went on to do great work. But first, it had to ride out the ARCO crisis. No one saw it coming, although perhaps they should have; but like many developments in the world of business and politics, the chain of events is often obvious only in hindsight.

Runaway Inflation

Conditions in the marketplace became more and more stressful. As Collyer put it:

We were severely impacted [by] an unprecedented worldwide demand for engineering, steel, and manufacturing capacity. . . . Prices for [all the materials and equipment that Syncrude was trying to order] became "price in effect at time of delivery" [and that continually drove up project costs and economics]. We were purchasing four draglines, the largest in the world [and there were only two manufacturers]. They quoted us a "cost in effect at time of delivery" [which proved ultimately to be $45 million apiece], and at that time, we did not even have a go [from the owners] to release the order. Bechtel's worldwide purchasing power was key to helping protect our delivery commitments through world price escalation.

Translation: The price of almost everything Syncrude needed to buy, skyrocketed. And so did the rolling estimate of what the project would cost.

Collyer's Strategic Contribution

In megaproject management as in war, success sometimes hinges on smart strategic decisions made before the fighting even starts. In 1973, looking into the future, Collyer had foreseen an issue that had the potential to sink the whole enterprise: the need for construction labour peace. Syncrude would need to be built with union labour. The overwhelming majority of the thousands of skilled tradespeople needed to construct the Syncrude complex would come from the hiring halls of Alberta's 21 construction unions. Long-term construction projects are highly vulnerable to union labour disruptions.

Any number of potential work stoppages during the construction period could've triggered a falling-domino effect that would've torpedoed the economic viability of the whole Syncrude enterprise.

And it wouldn't have to be a work stoppage at Syncrude. A stoppage at another construction project could've spread like a brushfire to engulf Syncrude too. Syncrude needed labour peace over the five-year construction period, and in 1974, neither the construction industry nor the skilled trade unions were prepared to guarantee it.

In Collyer's mind, Syncrude's owners didn't seem to particularly understand this problem, nor did many of the province's construction labour unions and various construction associations. Collyer said,

> Construction was badly fragmented and highly adversarial, generally trade-specific with no effective central voice [of labor and employers] for addressing industrial or labour relations issues. . . . The bargaining relationship with Building Trades was frequently adversarial, resulting in multiple strikes over the peak summer construction season. [Worse yet, construction owners] maintained a hands-off policy, leaving industrial relations problems

totally to their contractors to solve, something I was determined needed to change for larger projects [in order] to allow Syncrude construction to proceed during any labour strikes at other sites without capitalizing on the temporary increased labour availability.

The need was a direct consequence of the failure of the collective bargaining process in Alberta, which had existed prior to Syncrude's megaproject and without which it would create intolerable financial risk to the participants in what was a high-risk investment. While this was longstanding, no effective effort had been made to address it. Without it, the project could never have proceeded.
Together with Bechtel's labour relations manager, I wanted to create a positive relationship for Syncrude with the industry, which would be critical . . . to demonstrate our willingness to confront long-standing problems while helping ensure labour stability for the SCL project.

With the backing of Alberta's deputy minister of labour, the president of the Building Trades Council, and Bechtel, Collyer was able to gain support for a provincially mandated no-strike, no-lockout agreement for the Syncrude site for the entire construction period.

> While the owners' negotiations with government led to the enabling legislation, it would be left to Bechtel's Labour Relations Manager and me to shape policies and practices which would govern the site agreement and by which we would be bound. Not only were we sensitive to minimizing the adverse impact on other construction projects in the province, the level of which was high at the time, but we were setting the groundwork for the ability to have the lengthy transition between the construction force of over 10,000 and Syncrude's non-union operating staff occur without jurisdictional conflict precipitating job shutdowns.

Collyer and Bechtel worked closely with the province's deputy minister of labour to create the Construction Industry Industrial Relations Council, the only four-party group comprising engineering, construction contractors, owners, and government to address long-standing problems like collective bargaining, jurisdiction, work disruption, and manpower availability.

An unstated issue was that unless Syncrude could achieve labour peace during construction, operating unions like the Oil, Chemical and

Atomic Workers could have an opportunity to recruit Syncrude's permanent employees and turn its Operations Department (with its almost 5,000 employees) into a union shop. For a variety of reasons, Syncrude didn't want that to happen, which was partially why management set out to plan a radically new kind of workplace culture. As we'll see in Chapter 8, this helped pave the way for Syncrude's successful implementation of a team management culture that offered employees a pathway to a workplace of security, respect, and room for personal growth without recourse to union representation.

"We established open communications, trust, and a clear understanding of what constituted construction and what constituted operations," said Collyer, "which were respected by both parties without jurisdictional conflict throughout the lengthy two-year transition from construction to operation."

What the project agreement did for Syncrude's investors was remove one significant risk factor. All the other risks, like whether the plant would operate reliably, remained.

In a way, Collyer saw all these initiatives as part of a larger trust-building process that'd ensure a healthy relationship among all the players in the Syncrude system, including government, unions, contractor companies, Syncrude employees, and even Syncrude's owners, all of whom lived in silos.

The ARCO pullout, when the project had barely begun, put design and construction on hold for nearly four months; Collyer and Bechtel weren't given the go-ahead to resume engineering and procurement until after the Winnipeg Agreement was signed and the definitive cost estimate for the project was validated by the owners. This hurry-up-and-wait period destabilized the project team: contractors were hesitant to commit to badly needed Fort McMurray infrastructure projects, and prospective Syncrude employees (holding job offer letters) were hesitant to report for work in Fort McMurray.

As Collyer put it, "We had to change our execution strategy on the fly, and only an incredible performance by the project team enabled completion to occur on schedule and essentially on budget—something not achieved on subsequent Oil Sands projects."

And so, finally, the Syncrude project charged ahead.

"Bechtel's world-wide purchasing power was key to helping protect our delivery commitments [in the face of] world price escalation," said Collyer, meaning Bechtel may have been controversial, but it delivered.

It was one of the biggest and most complex design and construction projects in Canadian history. More than a million tonnes of material, vessels, and plant components were trucked and shipped by rail across North America (often after being shipped across the Atlantic and Pacific) to Fort McMurray. Some of these shipments were so large that they stressed the province's highways and bridges to their limits (Canada's railways had to figure out how to ship some of the larger vessels through the mountain tunnels).

John A. Lynn, Collyer's site construction manager, later said, "I still shudder when I think of some of the requests that went out to our procurement people: 'We need another five 200-tonne cranes—tomorrow.' Some of us had never even seen a 200-tonne crane. Somehow Bechtel would scour the countryside and somehow come up with them."

The foundations for the fluid cokers and the utilities plant were poured under huge, air-supported heated structures to protect the concrete and the workers from winter temperatures. A 600-foot smokestack was built, the tallest structure within 1,000 miles.

Some of the incidents that project managers feared did happen, like the time a train derailed in rural Saskatchewan and spilled 30 carloads of dragline components across a frozen wheatfield. But some didn't happen. A contract for the fabrication of two high-pressure vessels—monster steel contraptions manufactured only in three places in the world—was given to a factory in northern Italy whose unions were controlled by the Italian Communist Party. Some of Bechtel's more right-wing executives were extremely dubious about this arrangement, to put it mildly; but to the surprise of many, the Italians delivered the goods exactly on time and on budget.

Syncrude's own project management staff—outnumbered by Bechtel 10 to 1 at project sites—scrambled to stay on top of what they called "the well-oiled Bechtel machine."

It took five years and Herculean efforts, but by the end in the summer of 1978, everything had been brought together. Oil had been brought in by pipeline to get the fluid cokers started. The draglines and bucket-wheel reclaimers had been assembled and were digging. Plant managers were finding their feet. Newly painted equipment was starting to run. Initial batches of oil sand had been sent to the extraction plant for processing. The utilities plant was generating electricity and steam. Instruments had come to life. Lund's operating crews had been put through their paces and were familiarizing themselves with their equipment. The plant was ready for its official opening in early September —the sky would soon be filling with airplanes bringing VIPs and synthetic crude would be filling the pipeline.

The costs were astronomical. The original capital estimate in 1972 was $800 million ($5.8 billion in 2020 dollars). The final price tag was $2.1 billion ($12.5 billion in 2020 dollars). If it hadn't been for the fact that oil prices had risen commensurately, the inflation might well have doomed any further investment in the oil sands.

There would be glitches and very rough patches ahead, like the 1981 fire that badly damaged half the plant's coking capacity. But for now, the place was shiny and promising. On the morning of the official opening, I walked the newly painted and landscaped site with Brent Scott and remarked how finished and well-oiled it all looked. Scott said, "Plants always look this neat and clean and efficient before they have to actually operate. There'll be problems in start-up, don't worry. There always are."

This proved prophetic.

Even so, I had to think that William Van Horne should have been there, as he was at the driving of the last spike on the CPR at Craigellachie, B.C. If he had been, no doubt he would've said, as he did at Craigellachie, "The work has been done very well . . . in every way."

The big, tough, expensive job was Syncrude's first signature achievement and the one remembered today by the largest number of people.

We'll now turn to two other accomplishments that are more or less unrecognized: the economic liberation of Indigenous peoples in northeastern Alberta and the creation of Canada's first transformational workplace.

7. A PLACE IN THE SUN FOR INDIGENOUS PEOPLES

It's generally believed that the First Peoples of northeastern Alberta came from eastern Asia by land and possibly by boat between 12,000 and 15,000 years ago.[13] By various paths, small groups of them explored the Americas. At least 8,000 years ago, they began to populate the cold northern boreal forests that had taken root as the last ice age retreated.

It was a harsh and difficult land on which to live, but they were a tough and inventive people, and they figured out how to feed, clothe, house, and heal themselves using the animals and plants that the land provided. They were skilled hunters, gatherers, and fishers, and in time, they became skilled trappers and suppliers to the fur traders.

When the first European fur traders and explorers came to the northeast in the 1700s, the First Peoples had formed tribal groups called Woodlands Cree and Chipewyan. Their territories were roughly within the lands enclosed by Treaty 8 in northern Saskatchewan, Alberta, and northeastern British Columbia. Collectively, they're now referred to as "Indigenous peoples." During the 1960s and 1970s the accepted term was "Native people." A perfectly good phrase, it came to be replaced by the current term "Indigenous peoples," so that's the term we'll use.

In the late 1700s—according to legend, about nine months after the arrival of the first *coureurs de bois*—Indigenous peoples were joined by a mixed-blood people, the Metis. Together, we'll refer to them as the Woodland peoples.

For over 1,000 generations, with tenacity and ingenuity, the Woodland peoples survived and very slowly grew their numbers. In other parts of the world, empires rose and fell, and vast tracts of human history were written. But in the northeast, for a very long time, the seasons came and went, and little in the Woodland peoples' lives changed. They hunted and fished and gathered wild plants and medicines from the forests and lakes, and when they paddled down the rivers, they could see the thick black oil oozing from the high riverbanks. They mixed it with spruce gum to caulk their birchbark canoes, but other than that, it gave them little.

That, as we know, would change in the time of the tar sands.

* * *

In the early 20th century, eight generations after the trader and explorer Peter Pond came through their country, the Woodland peoples began to get work from the Europeans: hard manual work at first, hauling barges up and down the Athabasca River, and then as crews on the riverboats, where their knowledge of the river's sandbars was valuable. Some of them became riverboat captains. Some of them got seasonal work on the rickety Northern Alberta Railway that wound through the northeast until it reached Waterways, south of Fort McMurray.

In the first and second world wars, many Indigenous men volunteered for military service, performed well, shared the trauma of combat, and, like many whites, received little care from their country afterward. In the case of treaty Indigenous peoples, their treatment was worse: they weren't even given the right to vote until 1949.

Some of the Woodland peoples moved into the few small towns of the northeast, but by the early 1960s, the industrial world had brought them limited benefits: rudimentary public health, basic education, outboard motors to replace paddles, and skidoos to replace dog teams. They lived, many of them, in a marginal economy on the fringes of the fringes of urbanized, industrialized life. Some of them collected meagre cheques for treaty payments ($5 a year for most people) or provincial social assistance. Most of the modern economy's opportunities for personal growth, education, and wealth were unavailable to them. Whether they wanted to be or not, they were poor.

Social Credit Rediscovers Its Reform Roots

In the 1960s, social, economic, and political change began to sweep the Western world. In Africa and Asia, anti-colonial revolutions were underway. The winds of change were everywhere. The Vietnam War—and opposition to it—began to fixate popular society in North America. In the United States, the civil rights movement began to drive political change. Student revolutionaries roiled college campuses. In 1968, students led by Daniel Cohn-Bendit captured the downtown streets of Paris and came within a whisker of overturning the French Fifth Republic.

In the mid-1960s, these perturbations began to be felt in a place the Canadian political class would've described as the unlikeliest of all: in the province of Alberta and its Social Credit government.

Social Credit, a populist movement, was born in the Depression as a reform party, but it had been in power for 30 years and had become set in its ways. Then Premier Ernest Manning's son Preston and a graduate sociology student named Erick Schmidt—the latter an advisor to cabinet—set about letting in some fresh air. Manning and Schmidt began to promote progressive policy thinking. Their watchwords were "human resources development." And the first human resources they decided to target were the province's Indigenous peoples, particularly in the north.[14]

Here and there in Alberta society—in politics, the media, business, and the community—there was a growing belief that something needed to be done about the plights of Indigenous peoples. Partly in response to this feeling, the Alberta government created a community development program aimed at accelerating the process of social change by animating neglected Indigenous communities to demand it.

The Community Development Branch was headed by a bright, mischievous community organizer named Jim Whitford. Whitford, who cheerfully described himself as a "strategic shit-disturber," was a student of the radical community organizer Saul Alinsky. Years before the rise of the civil rights movement, Alinsky had organized successful community uprisings in Rochester, New York, where he mobilized the Black community against the Eastman Kodak Company. Whitford intended to apply that kind of community organizing to depressed communities in Alberta.

He and his number two, an ex-Indian Affairs official named Ben Baich, believed that 100 years of living as wards in a paternalistic system—today it would be called "colonial"—had made Indigenous peoples not just poor but passive and resigned to poverty. Whitford, Baich, and their colleagues were a merry band of agitators who believed that if trained community animators could be placed in Indigenous communities across northern Alberta, then economic, cultural, and social change would be accelerated and a new generation of Indigenous leaders would shake up depressed Indigenous communities.

If the idea of civil servants promoting radical social change strikes you as odd, you're not alone; many orthodox Albertan civil servants in the day were scandalized by Whitford "and that bunch." But Whitford had political cover in the form of cabinet support, so he got away with it.

An invisible network of players who wanted to bring Indigenous peoples into the mainstream of economic life was starting to come together. One node in the network was a federal-provincial adult education program for Indigenous peoples called Alberta NewStart. The principal of NewStart was an ex-Canadian Forces warrant officer named Jack Shields, a smart, ambitious operator who made connections in the education and business communities. Shields knew how to get things done. He and Frank Spragins spoke the same language. Shields went on to enter politics and was elected to three terms as a conservative member of Parliament for northeastern Alberta.

In a strange convergence of events and personalities, one of Whitford's senior community development officers, an ex-Mountie named Terry Garvin, joined Syncrude in 1973 and helped to precipitate the company's Indigenous peoples' revolution. We'll return to him in a moment.

Far-reaching changes in the relationship between mainstream white society and Indigenous peoples in northeastern Alberta, when they came, didn't just happen—they were precipitated by an executive with a vision of how the industry could and should change people's lives, an activist who spread the vision and built relationships, and an organizer who could translate the vision into a corporate enterprise to recruit and train Indigenous peoples (and company managers) and purchase services from Indigenous-owned enterprises.

What follows is a description of how Syncrude's Indigenous peoples program blossomed from Spragins's vision into a full-bodied corporate initiative with serious funding and serious results. This process revolved around two people. One man, a change agent, old it to northeastern Indigenous communities and to Syncrude upper and middle management. He was Terry Garvin. A second man, an organizer, built what was then called Syncrude Native Affairs into a muscular corporate program with a staff, a budget, and a mandate; and he got large numbers of Indigenous peoples into jobs and careers. His name was Alex Gordon. Their initiatives—and the response to them by thousands of Indigenous and other peoples—made it possible for Syncrude to become the earliest and largest employer of Indigenous peoples in the history of Northern Canada.

Here are some glimpses of how it happened.

The Vision

By the mid-1960s, small and isolated Indigenous communities in northeastern Alberta—places with names like Fort Chipewyan, Fort McKay, Janvier, Conklin, Anzac, and Kikino—were still mired in poverty, unemployment, and life on social assistance. The private sector was bringing them little benefit. GCOS had been built but was struggling and, in any case, uninterested in reaching out to Indigenous communities; otherwise, an "oil sands industry" was more an idea than a reality.

In 1965, Syncrude named Frank Spragins as its president. Spragins was determined that Syncrude wouldn't only build the first full-scale oil sands project but would also use oil sands development to help lift Indigenous peoples out of poverty and give them a leg up in the modern industrial economy.

In 1966, Syncrude had fewer than 100 employees and little physical presence in the oil sands region. It wasn't yet in a position to start creating jobs for Indigenous peoples. Even so, Spragins put down a marker for the company's intentions and began meeting with Indigenous leaders like Sam Sinclair of the Alberta Metis Federation to explore how jobs *could* be created for Indigenous peoples.

Around 1968, Spragins and Preston Manning put together a *very* low-profile advocacy group, so low-profile it was almost invisible. It carried a deliberately stunningly uninteresting title: The Economic Development Discussion Group. Its purpose was to identify opportunities to hire Indigenous peoples in the oil, gas, and utilities industries. At different times, up to 20 major businesses participated along with provincial and federal civil servants. Attendees included Syncrude, Suncor, Esso Resources, Shell, Mobil, Husky, Dome, and TransAlta Utilities.

The group met sporadically until the early 1990s. Its minutes document dozens of small hiring initiatives and community development projects across northern Alberta. Although no formal measurement of its job creation appears to have been done, it appears its members probably hired a few hundred Indigenous peoples into seasonal or permanent jobs.

Spragins was determined that when Syncrude got the go-ahead to build, it would become a major employer of Indigenous peoples. He communicated his vision to a cross-section of Indigenous, business, professional, and government organizations, and it was widely understood. To my knowledge, it was never stated as a quota or fixed percentage of the workforce but rather as a sincere commitment of best efforts backed up by an understanding of the cultural obstacles that would need to be addressed within both business and Indigenous communities.

Preston Manning remembers:

> Frank told me the story of recruiting one of the first groups of Fort McMurray/Fort Chipewyan natives for one of Syncrude's first programs to train them as heavy equipment operators. The group got through the training and were given their first on-the-job assignments under the direction of a crusty, old-school foreman. One of the trainees had learned on a John Deere machine but was assigned to a Caterpillar machine and couldn't get it started. The old foreman reamed him out in front of the whole crew, in essence destroying the frail confidence of the native fellows that

they were fit to join the white work force—a confidence which had been painstakingly built up by the training program. Frank's conclusion was that Syncrude not only needed to train the native folk but also train its foremen and managers on how to relate to them, positively and constructively.

Frank . . . told me the story of the necessity of thoroughly understanding the native cultures and languages if one was going to make much progress in involving them in a project. The example he related was that in the Chipewyan language, there is apparently no such thing as the imperative tense of the verb. . . . In that language, you can't give an order to someone by saying "You go and do this!" The way you give an order in Chip is to say "Why don't we go and do that?" And this suggestion gets its force as an imperative by who says it. . . . He felt it was very important for this to be understood by white management giving orders to Chip workers—and another reason for including training for management if they are going to be called upon to manage native workers. [Frank may have learned this] from Ted Van Dyke Jr. . . . who was a trained anthropologist/sociologist who immersed himself for seven years in the Fort Mackay community.

Why did Spragins embrace the cause of creating opportunities for Indigenous people? It's an important question but a difficult one for a historian to answer. He wasn't a man who speak freely of his feelings. Looking back on our time together, I did not gain any psychological understanding of why he adopted the cause of opportunity for Indigenous peoples. I'm tempted to wonder whether his Depression upbringing in Mississippi and Texas—and his childhood experiences in a South of severe Jim Crow laws and segregation—somehow gave him a sympathy for the underdog. All I know is that even later in life, even when he was struggling with cancer, he never spoke about it.

Social Responsibility

Spragins was the first to understand that *how* the oil sands should be developed was as important as *whether* they should be developed. What kind of industry would the oil sands be? How would that industry relate to the people of the northeast, the people of Alberta, and the people of Canada? How would it treat the environment? In the 1960s and early 1970s, the term and concept of corporate social responsibility hadn't yet entered popular parlance. But it was very much what Spragins was talking about.

In 1973, after receiving a green light from the Energy Resources Conservation Board and negotiating a realistic royalty agreement with Premier Lougheed, Syncrude finally began to forge ahead. But it did so with no detailed blueprint for an Indigenous opportunities program. No policies had been drafted, and no expert staff had been hired. All that was still to be done.

Syncrude's Indigenous affairs program took almost two decades to reach flower. It was the work of many people, not just the three who led it. I acknowledge the contribution of many, many others and have listed some of them in the appendices.

The Change Agent

In 1973, it was time for Syncrude to start translating Spragins's vision and sentiments into actions to recruit and train Indigenous peoples. In other words, start putting boots on the ground. But whose boots, which ground, and when?

I believed that translating Spragins's vision into reality would need a company change agent on the scene in Fort McMurray.[15] I had in mind someone who really *knew* Indigenous peoples in the north and was trusted by them. Someone who could prod Syncrude's managers to identify opportunities for Indigenous peoples, hire them, and provide cultural sensitivity training for company managers. Most importantly, someone who could build a network of relationships between Syncrude's local management in Fort McMurray and local community leaders across the northeast.

It would require skill and time. Plant construction was just beginning. The Projects department was just beginning to get its arms around its staggering task. The Operations department, which would eventually employ more than 4,000 people, was just starting to staff up. Getting senior and middle management's attention to the challenge of Indigenous employment would be like jumping onto a parade float and persuading the driver to turn down a different street.

The man I knew with the right combination of commitment, firmness, and tact was Terry Garvin.

Left, Terry Garvin in the 1970s. (Photo courtesy of Syncrude). *Right*, Terry Garvin around 2015. (Photo @Warren Harbeck).

Garvin was born in 1930 and grew up on a farm near Craik, a small town in south-central Saskatchewan. In 1951, he joined the

Royal Canadian Mounted Police and spent the next 13 years posted in western farm towns and Indigenous communities across northern Saskatchewan (Uranium City), northern Alberta, and the Northwest Territories. If you were filming *Sergeant Preston of the Yukon* and were casting the Sergeant, you'd have asked central casting for an actor who looked like Garvin. He was a broad-shouldered weightlifter who had mushed dogs across the Northwest Territories, hunted caribou with the Dogrib Indigenous out of Fort Rae, and cultivated deep and lasting friendships with dozens of Indigenous people and their families across the northeast.

During his RCMP career, Garvin learned to speak Cree and developed a deep and abiding love and respect for those he later called the "Bush Land People."[16]

During his police days in northern communities, the Mountie was expected to be everything from police officer to magistrate to social worker to community recreation officer. While Terry was deeply committed to social change, he was no academic bleeding heart: he was well acquainted with the gritty realities of the real world. One day, when he was still an RCMP constable in Bonneville, Alberta, he was threatened by Steve Holowaty, leader of a notorious rural Albertan gang. Garvin strode to Holowaty's car, seized him by the scruff of the neck, smiled, and said, "Steve, you need to understand that I'll never let you hurt me."

I ran across Garvin in 1966 when I was writing about community development for *The Edmonton Journal.* I liked him immediately. We later met several times when I was legislative assistant to the Alberta minister of education. I knew about and respected his passionate devotion to the welfare of Indigenous peoples.

Garvin's assignment was to build relationships and trust between Syncrude and the Indigenous communities of northeastern Alberta and to work with Syncrude management in Fort McMurray to follow through on Spragins's commitments to maximize opportunities for Indigenous peoples. In other words, to be a catalyst for change.

He wasn't a perfect fit for the job (and the job wasn't yet fully defined.) He'd never worked for private industry and didn't fully trust it. How, then, would he go about animating Syncrude managers or local Indigenous people? We knew he was the kind of guy who could gain the respect of hard-nosed engineers and mining bosses and bring people together. Management would have his back.

One white manager who was, shall we say, skeptical of hiring Indigenous workers told Garvin that "Indians aren't reliable employees. They think they can just go off and hunt moose whenever they feel like it." Garvin's response: "None of the native communities in the North have ever had a reliable *employer*." How could they be expected to understand the concept of industrial job discipline?

It didn't hurt that he'd been a Mountie. When I told the elderly (white) ex-mayor of Fort McMurray that Syncrude had hired Garvin, he nodded and said, approvingly, "Yes, he was with the Mounted Police, you know."

Two Groups that Didn't Make a Difference—and One that Did

You might imagine that Syncrude became a leading employer of Indigenous peoples because government required it or because the company yielded to pressure from Indigenous organizations. Neither happens to be true.

During the period covered by this book, there *were* no federal or Albertan government policies or guidelines related to the hiring of Indigenous peoples. Affirmative action guidelines were still well into the future. Premier Lougheed, who was in office during the scope of this story, had little interest in the whole subject. On taking power, one of his first initiatives was to abolish the civil service organizations (the Human Resources Development Authority and the Human Resources Research Council) that the outgoing Social Credit government had created to advance issues like Indigenous hiring and training.

As to pressure from Indigenous organizations,

the only group that played any significant role with Syncrude was Sinclair's Metis Association of Alberta. Sinclair knew Spragins well and worked with him.

The much larger and better-financed Indian Association of Alberta (as it was called at the time) had little interest in corporate action. Its president, Harold Cardinal, was focused chiefly on pressuring the federal government to honour its treaty commitments.

In 1976, after Syncrude's Indigenous hiring program was already well established, the Indian Association of Alberta awoke to the issue and loudly called for Syncrude to commit to hiring and training Indigenous peoples, which it was already doing. The subsequent agreement between Syncrude, the association, and the federal government simply papered programs and procedures that Syncrude had already put in place. The association and the feds then went back to their offices and were not heard from again.

First Encounter of the Close Kind: The Fur Trapper Negotiations

Early in 1973, Garvin reported a situation that would be the first test of Syncrude's corporate social responsibility. Bulldozers were stripping the trees and soil from a six square-mile rectangle of forest land that would become the site of the project. This site clearing would obliterate part or all of the livelihood of three Indigenous trappers..

We met the three men in Fort McKay and talked about their traplines and their lives. We explained Syncrude's plans. Terry and I both felt that the men should be compensated fairly; management supported us. But how to work out a settlement wasn't clear. Trapping fell under provincial jurisdiction constitutionally, and at the time, there were no provincial government requirements or guidelines for trapper compensation. In fact, when we sounded out officials of the Alberta Ministry of Lands and Forests, which was responsible for licensing fur traplines, we were told that no compensation was needed: a trapline license belonged to the crown and conveyed no ownership rights to the trapper; the crown took no responsibility for finding a replacement trapline for any trapper thus displaced. In fact, the civil servant in question expressed mild annoyance that we were even thinking of doing this, since it might screw up his administration.

The government's policy struck us as unjust and unfair, and we set about researching how Indi The government's policy struck us as unjust and unfair, and we set about researching how Indigenous peoples in other parts of Canada were being compensated for their displacement by resources projects. The only current examples that we could find were the James Bay land settlement of Hydro-Quebec and agreements that BC Hydro had worked out for native people whose lands had been flooded by hydroelectric dams.

We interviewed both crown corporations and then invented a rough-and-ready compensation formula that would provide each of the three men with a cash settlement equal to 10 years' income from their traplines. The three men were all in their fifties and within 10 years of normal retirement.

The trappers (and their legal counsel) agreed. In fact, the trappers were surprised that the company was offering them anything. Establishing their usual annual incomes from trapping proved challenging. None of the men kept financial records, and in any case, trappers' incomes often fluctuate year to year due to shifts in fur prices and animal populations. We asked them to estimate their average trapping income, rounded it up to the nearest $1,000, and took them at their word. With the oversight of their counsel, they signed settlement agreements and we cut them cheques.

Our doctrine of corporate social responsibility had, we thought, passed its first test. Its next test would be more serious.

The Bechtel Two Step

In an earlier chapter, we recounted a roadblock to Syncrude's Indigenous hiring goals in the form of opposition from Bechtel's first construction manager.

It was clear that at least one Bechtel senior manager didn't understand Syncrude's commitment (or thought it was corporate baloney) .This had to be confronted, so Brent Scott, then Syncrude's executive VP, called for a project-review meeting with Bechtel senior management at their San Francisco headquarters.

I remember the occasion vividly. It took place in a Bechtel conference room high over the city. A summer fog was rolling in from the bay; we could see it slowly climbing up the side of the office tower across the street. Finally, it rose to envelop the conference room windows.

Scott opened the meeting by explaining Syncrude's objectives and community commitments. A Bechtel labour relations person flown in from the site defended Bechtel's record in Indigenous hiring. The next week, Bechtel appointed a Canadian engineer to replace Mister C. Whether he was fired or simply reassigned somewhere else isn't known.

While the meeting got Bechtel to support Syncrude's Indigenous hiring goal, it is doubtful that a large number of Indigenous people got hired into construction jobs as a result. The reason lay in the historic hiring practices of the construction industry, its unionized section in particular.

Because Bechtel was a party to the no-strike, no-lockout site agreement between Syncrude and the 21 construction trade unions and because, under the agreement, all construction workers had to be hired through union hiring halls, most of them based in Edmonton, and because most construction unions at the time had few Indigenous members, Bechtel could not have promised to hire any fixed number of Indigenous workers. It's conceivable that some of the northern Alberta members of some of the unions—the labourers' union especially—were Indigenous people and that some number of these workers were included in the calls that went out from the union halls. But to the best of my knowledge, no records were kept, and in the 1970s, it was probably considered a violation of human rights legislation to stipulate a worker's racial or ethnic identification in the first place.

Given that, at the peak of construction, there were in the vicinity of 8,000 unionized construction workers on site, it's conceivable that perhaps a hundred Indigenous workers were employed at the site, mostly as labourers. By late 1974, it was apparent that if a large number of Indigenous people were to get oil sands jobs, the jobs would have to be in plant operations, not construction. How that happened is described below in the story of Alex Gordon.

Two other aspects of Garvin's outreach to

Indigenous communities bore greater fruit: interest in gaining employment with Syncrude and the contracting of services from Indigenous companies.

The Benefits of a Human Face

Syncrude wasn't known or trusted by Indigenous peoples in northeastern Alberta's communities, and its job opportunities and requirements were little understood there. Plus, few Syncrude managers had any sense of the outlying communities or the potential of their people. The people of the region were far flung and varied. The critical first part of the relationship-building process with Indigenous communities, therefore, was putting a human face on both the company and the community.

Garvin launched this people-to-people process by organizing visits of Syncrude management to Indigenous communities and community leaders to the Syncrude site. Each side briefed the other in these meetings, and everyone broke bread together.

Syncrude managers would need to learn about the cultural factors, and Indigenous communities would need to get to know Syncrude managers and something about the Syncrude enterprise.

The Indigenous people of the northeast spoke three languages—Chipewyan, Cree, and Dogrib —as well as English and some French. The only Indigenous community near the Syncrude site was Fort McKay, 16 kilometres down the road. The other communities of the northeast were Fort Chipewyan, 250 kilometres north of Syncrude and, in the 1970s reachable only by aircraft and sometimes in the winter by ice road. In the south, there were Conklin, Anzac, Lac la Biche, and Kikino (the latter 212 kilometres from Fort McMurray). Few of the Indigenous peoples in the northeast had ever worked for a modern industrial employer (the exception being the Metis, who worked on the Athabasca River for Northern Transportation).

These community and site meetings built respect and trust, and they also taught Syncrude executives that for community leaders, sending young people off to a distant employer was a two-edged knife. As Neil Lund, the company's senior operations VP, remembered:

> The problem as seen by the Elders, or at least the chiefs—chiefs and Elders are two different categories—was that they really didn't want to lose their young people, which was of course who we wanted to recruit. We weren't recruiting people who were about to retire. And of course, we'd also be after the best-educated people. [The chiefs and Elders] were somewhat reluctant.[17]

Contracting Opportunities

Thinking about job creation led to a realization that not all jobs had to be permanent positions within Syncrude; if local people could form companies to deliver services, Syncrude could contract with those companies and create large numbers of indirect jobs. During Garvin's time, three such companies—a dry cleaning plant at Goodfish Lake, a clothing repair service at Fort McKay, and a labour exchange in Fort Chipewyan—found their first footing.

Goodfish Lake was an Indigenous community several hundred kilometres south of Fort McMurray. When its chief expressed an interest in getting an enterprise started, Syncrude helped the band set up a dry cleaning plant and laundry service, finding a retired CESO executive in Quebec who agreed to train the employees and oversee the setting up of the plant. Syncrude gave the Goodfish Lake people a contract to pick up and clean the large amounts of soiled industrial clothing and accoutrements that the plant turned over every day. The Goodfish Lake laundry and dry cleaning plant became a significant enterprise that is still in operation 30 years later.

The Fort Chipewyan labour exchange was built on an agreement that the local community would find teams of workers that Syncrude would fly to the plant and house in the camp for seven-day shifts. That program generated a significant flow of income into Fort Chipewyan without disrupting community life or requiring families to leave the community and move 250 kilometres away to a strange city.

The contracting program was born during Garvin's time and really caught fire during Alex Gordon's time and later. By the mid-1980s, it had exploded, so much so that Syncrude came to be known as the largest private-sector employer of Indigenous people in Canada and the largest user of Indigenous-owned enterprises.

Now flash back to 1977. The construction of Syncrude was entering its final year. One to two hundred Indigenous people had found either direct jobs with Syncrude or indirect jobs with contractors. But it was clear to Spragins that something more would be needed if the employment of Indigenous peoples was to realize its full potential.

A big part of the problem was that, in an organizational sense, nobody in Syncrude management completely owned the Indigenous employment program. Garvin was still educating managers and building ties between the company and Indigenous communities, and the company's Indigenous employment goal was still supported by staff people in the human resources department on the operations side, but the program was growing very slowly, if at all. Indigenous affairs needed a *driver*. A tough, smart, totally dedicated manager who could coax, persuade, and, when necessary, pound the table to make things happen.

Spragins knew such a man, and that man was Gordon.

The Organizer

Left, Alex Gordon in 1978. (Photo courtesy of Syncrude.) *Right*, Alex Gordon in 2018. (Photo courtesy of Alex Gordon.)

Early in his life, Gordon established that he was a can-do, bootstraps kind of guy.

A Metis who grew up on a farm east of Edmonton, he got his first job in 1958 as the diesel engine man on a tugboat pulling barges up the Slave River from Fort Fitzgerald, Northwest Territories. He worked his way up from there, doing everything from skinning wood bison in a northern slaughterhouse to head-manning a survey crew with a rod man named Cigar Mercredi to survey the new townsite of Pine Point.

In the course of a 30-year career in the territorial government—with time out at age 29 to get an honours commerce degree at the University of Alberta—Gordon did everything from learning to speak Inuit to working on the *CD Howe* medical ship visiting all the eastern Arctic outposts. He travelled to Inuit camps by dog team and learned how to build igloos. He was the area administrator responsible for Cambridge Bay on Victoria Island in the High Arctic. He took community development training from the Peace Corps and developed guidelines for the McKenzie Valley Pipeline (which was never built). By 1977, he'd risen to director of planning and program evaluation, reporting to the commissioner of the Northwest Territories.

In 1977, Syncrude was in its third year of construction and nearing start-up. Spragins had followed the progress of the Indigenous affairs program closely. He knew it was high-centred (a backcountry expression for a vehicle stuck in deep ruts in the mud). It needed a chief ramrod, somebody who could get things done in a large organization.[18]

Spragins had known Gordon for some time. Gordon had been friends with Spragins's son Rob at the University of Alberta. Spragins had followed Gordon's career in the Northwest Territories; Gordon had even flown him in once for a fishing trip near Yellowknife. Spragins figured it was time to bring Gordon into the picture. He organized a dinner at the Edmonton Petroleum Club with Alex and Carl Sherman, the number-two person in Syncrude Operations.

Sherman, a hearty Cape Bretoner who'd managed oil refineries in the Maritimes, was a perceptive man; he was impressed by Gordon's CV and obvious talent. During negotiations over salary and emoluments, it was agreed that Gordon would be given full responsibility for a Native Affairs Department reporting to Sherman and, through him, to the executive committee and the owners. All Indigenous staff would be transferred to Gordon's department, which would have offices both at the plant (still nine months from commencing operation) and in Fort McMurray.

On the job, Gordon went about evaluating the Indigenous program in the good, systematic way he had learned in the public service.

> I did a full review of the outcomes of the program, listened to staff members about the program, looked at targets, issues and concerns and asked for team member suggestions on how we could improve our programs. We did a study of Native turnover to determine why native people left the company. Some left without notifying their supervisor. The turnover rate was at 30 percent. We sent two staff members to interview each person who had quit. Our staff was fluent in Cree and Chipewyan and was able to speak to them in their own

language or in English. We received a 100 percent response.

Challenges and Obstacles

It was apparent to Gordon that the two biggest barriers to hiring and retaining more Indigenous people were twofold. First, supervisors didn't fully understand or respect the cultural differences of Indigenous employees, which was creating "resentment, distrust, and poor communication." Second, pockets of bias, "even in an organization that espouses a welcoming work environment," were causing tensions. Says Gordon,

> Some Natives were too shy to confront supervisors and co-workers when things were not going well at the workplace. And there was racism in some pockets of the organization, sometimes in the form of managers who did not want to have any Indigenous people in their department. One manager had to be transferred to an advisor role because of his negative attitude toward Indigenous people.
>
> A significant problem was education; most companies required new hires with at least high school education or GED. The graduation rate for aboriginal people in the 1970s was around 12

percent. Drop out rates for high school were over 80 percent.

There were a number of misconceptions by supervisors and co-workers that Indigenous people were given special treatment. Some saw the Syncrude Native program as a means for natives to get a free ticket to a job. And some single mothers with qualifications could not leave reserves for lack of day care in their community.

Gordon's Strategy

Gordon acknowledges that the support he received from Syncrude management played a major role in the success of his program. Scott, who was CEO during Gordon's tenure, was a solid champion of the program. Dennis Love, who was general manager of mining, was a major booster. Says Gordon,

> Dennis took the lead on hiring and training aboriginal people. He was convinced that native people had the potential to aspire to any job in the mine, be it an operator for heavy equipment or for administrative staff. In Denis Love, we had a man with fire in the belly and a strong leader who we used to exert influence within the department and with his peers. Other operational managers also provided our team with strong targets.

It was our job to meet these expectations. My goal was to bring in successful people who had highly developed skills in their field of work. We recruited throughout the province, and in some cases, we found Indigenous people in other provinces who had demonstrated great capability in their field. We could demonstrate to managers that native people could be reliable at work every day and stick to the job.

In a short time, Gordon grew his Indigenous affairs team to 12 professional Indigenous staff members. Several of them became recruiters and went to every corner of Alberta and beyond. Hiring and training targets were set based on the ratio of the number of Indigenous and non-Indigenous populations in the northeast region. "This meant that 8 percent of our workforce would be a native. With a workforce of 5,000, we would recruit at least 300 people."

Eventually, Syncrude was able to do even better than that. In 2018, the company reported that in the previous five years, Indigenous people made up 16 percent of its workforce of 5,600 people.

Each operations department was urged to set hiring and training goals where vacancies existed. In some cases, a few departments set no limit on the number of recruits the company could bring on board. Gordon looked for champions who'd be willing to step up and hire more Indigenous people.

The company provided letters of intent to candidates that didn't have Grade 12. When they demonstrated completion in the GED program at a local college, they were guaranteed a job as trainee operator, apprentice, or administrator. This worked well. The Indigenous cultural program was also expanded. Dr. Ted Van Dyke was brought in as an instructor for cross-cultural training. This program was mandatory for all new [Operations?] hires so that they could gain a better appreciation of Indigenous peoples, their histories, their cultures, and the problems they faced in a changing world.

Gordon met with chiefs and Elders of First Nations across the province. This built support but also uncovered problems:

> Most were supportive of our program, [but] some said that if you take our people off reserve, we'll have fewer able-bodied people left. So we held recruitment nights on reserve and in nearby communities or, in some cases, the closest to a reserve where we could get a boardroom for meeting. We recruited some people from southern (Blackfoot and Stony) reserves; however, the turnover was high since they were Plains native people who didn't like living in the forest. They were lonesome for wide-open

prairies. We also tried pairing up a new hire with a friend, so we would hire two people hoping for mutual support. That didn't work; they still missed the south. So we chiefly zeroed in on central, eastern, western, and northwestern Alberta.

In Alberta, most Indigenous people living south of the Battle River are members of the Blackfoot Confederation, a Plains Indigenous nation.

Building the Fort Chipewyan fly-in program created special challenges. Gordon wanted to expand bitumen recovery using a pond skimmer developed by a Syncrude design engineer. The company envisaged flying in teams to work seven 12-hour shifts. That conflicted with the province's labour laws, so Gordon met with the minister of labour and arranged to fly three cabinet ministers to Fort Chipewyan in old DC3. They observed first-hand that unemployment was very high, people were still getting water by going to the Lake Athabasca with their pails, and Indigenous residents were still using outdoor toilets. They met with local leaders.

Following the trip, the government agreed to have the fly-in crew work 7–12 hour shifts for two crews of six people. This provided work for 12 families. The work ended in the fall so the men could receive their annual vacation and then work on the land trapping and hunting.

Gordon held community meetings to attract workers for the program.

"I was looking for people who had skills in running fire pumps to fight forest fires, savvy with motorboats and general mechanical aptitude," he said. "These were older people who were still living off the land trapping and hunting. We were successful in manning up the crews. . . . The men stayed at our camp."

And the program is still operating today.

Contracting Indigenous-Owned Enterprises

The dark horse in the race to create economic opportunities for Indigenous people turned out to be contracting services from Indigenous companies. The program had a slow start during Garvin's time, rapidly gained huge momentum during Gordon's term of office, and exploded in the mid-1980s with the creation of dozens of Indigenous-owned enterprises that negotiated contracts not just with Syncrude but with other oil and natural resources companies across the province.

Let's have a look at some of the early success stories that gave a boost to the whole process.

The Goodfish Lake Dry Cleaning Plant

Says Gordon about the plant,

> The first major contract was outsourcing our dry cleaning to Good

Fish Lake Reserve. Chief Sam Bull started the Goodfish Lake Business Corporation owned by the Whitefish Lake First Nation #128. His vision was to build a strong economic foundation that creates prosperity, employment for aboriginal people, and protect the environment. This five-year contract was in place when I arrived. However, there was still work to do to train workers on how to operate the plant. I was able to secure the help through CUSO to attract a retired dry cleaner owner from Quebec to live on the reserve to train the staff in dry cleaning. The chief set up a trailer for him and his wife. He performed the work without pay. He was able to train a crew of 15 or more women to run the plant. This plant was the most sophisticated up to date plant in the country. In order to deliver clothing to the Syncrude site, the reserve purchased a container van and hired drivers. Once other companies saw how successful they were in delivering a great product they engaged the Goodfish Lake Corporation to do it's dry cleaning. Since then, the reserve has built a larger 17,000-square-foot plant, started an industrial garment division, and owns a protective clothing retail store in Fort McMurray.

Fort McKay Seamstresses

Says Gordon about a group of seamstresses,

Since we were dry cleaning up to 70,000 clothing items consisting of coveralls, pants, shirts, parkas, protective clothing, and so on, we set up a sewing and mending service on site. This required purchasing sewing machines and equipment to run a sewing shop. Seamstresses would mend, sew on new nametags, and deliver the clothing to various lockers throughout the plant.

Gordon met with several women from Fort McKay to offer them jobs as seamstresses. They met in the grass outside one of the buildings on the reserve and discussed the opportunity. The women were worried about a possible explosion on site and would not readily commit to working at Syncrude, so he offered them a site tour so they could observe first hand what was going on with operations in the upgrading area. They were "quite impressed" with the operation and agreed to give the jobs a try. Syncrude hired older women who were seamstresses in their own right and who had experience sewing clothing, moccasins, and the like. The women suggested Syncrude hire some younger women who could train and learn with the guidance of the Elders, and this was done. The company hired 12 women and set them up in the building near the upgrading area. This program was very successful and is still in operation.

Neegan Enterprises

An Indigenous company, Neegan Enterprises, was given a contract to do road maintenance on site. They operated graders and Caterpillar tractors as well as other maintenance equipment. Joe Dion, chief of the Kehewin Reserve, was the president of the company. He went on to be the chairman and CEO of Frog Lake Energy Corporation, which extracts oil on a reserve 200 miles east of Edmonton.

Jitney Service Contract with Fort McKay First Nation

Syncrude needed a bus service to bring people from Fort McKay to and from work and a jitney service to take Syncrude personnel to and from the plant to Fort McMurray. The company worked with the Fort McKay First Nation Chief Jim Boucher to provide up to 15 passenger buses. Chief Boucher eventually formed The Fort McKay Group of Companies, which earned an average gross annual revenue of $1.7 billion over five years, according to a news release from the Government of Canada. His operations include earthworks, logistics, lodging, and catering, and he's one-third owner of the Suncor East Tank Farm. Fort McKay First Nation now experiences virtually zero unemployment.

Labour Crews

Syncrude's upgrading department needed extra labour to work in and around the plant. The company placed a contract with an Indigenous-owned firm from McLellan, Alberta, to provide 15 Indigenous labourers on a rotation basis. The contractor bussed people from the High Prairie and McLellan areas, and they stayed at Syncrude's camp. In order for the contractor to make payroll, Syncrude advanced funds for the first few months until the owner was able to cover the payroll with revenues.

Family Support

The Indigenous team of Mariella Sneddon, Pat Harp, Alvena Strasbourg, and Rita Martin were successful in developing and implementing a family counseling service to assist new Indigenous hires moving to Fort McMurray from out of region and within the Wood Buffalo region. They met regularly with new-hire families and helped them adjust to new housing, register their children for school, and take advantage of community services like banking, the friendship centre, and the college. This program helped families settle in to their new environment of city life and resolved the needs of new recruits and their families.

Epilogue

Forty years later, Gordon looks back with pride. "When I started, our native turnover rate was 30 percent," he recalls. "After two years, it was 6 percent, lower than other workers at the site. At the end of two years, through direct hires and vendors, we had hired 600 native people, the highest of any corporation in Canada." He acknowledges the role of his team in delivering "one of the most successful, recognized programs in the country."

In 2018, Syncrude announced that Indigenous peoples—workers self-declared as First Nations, Metis, or Inuit—represented 10 percent of its 4,700 full-time workforce. In the five years previous to that, Indigenous people represented 16 percent of the company's new hires.

And the company's contribution to the economic development of Indigenous communities didn't stop there. Between 1992 —when Syncrude began carefully counting the flow of funds—and 2018, the company spent $3 billion on contracts with Indigenous-owned companies. Today the company works with more than 50 Indigenous-owned companies in the region.

The liberation of Indigenous people was especially stunning in Fort McKay. In 1986, the Fort McKay Group of Companies began operations with six employees and a single janitorial contract. Today, the Fort McKay Group of Companies employs more than 1,450 people who work with oil sands companies across the region in everything from earthworks, logistics, site services, and environmental services to a range of oilfield construction and services.

Thus, beginning in the early 1970s, Syncrude powered a social revolution in northeastern Alberta. It was a revolution driven neither by a sense of charity nor the prodding of government but by a belief that corporate social responsibility meant doing the right thing because it was in the long-term best interests of society and business.

8. A NEW AND BETTER KIND OF WORKPLACE

It's one of the most powerful central organizing principles of the human race. It's the "fraternity" in "liberty, equality, and fraternity." It's the Band of Brothers and the U.S. Marine Corps leaving no wounded soldier behind. It's Athos, Porthos, Aramis, and D'Artagnan. All for one, and one for all. It's King Lyonidas and the Three Hundred facing the hordes of Persians at Thermopalae. It's "if we don't hang together, we'll all hang separately." It's the idea that all of us are one.

Powerful stuff. But how could anyone apply this principle to the life of the modern corporation?

Deep down, many of us would like our workplace to be something like a *family*, a nurturing place that cares for its members. But a corporation isn't a family, and some would say it's not meant to be. It's an enterprise designed for efficient production, not nurturing.

Still, can it not be more than that? An army, after all, is a structure designed to advance policy goals by killing people. But the most successful armies are more than that. They are a unity, a powerful bond of people who achieve more together than the sum of their individual efforts. So could not a corporation become something more? Could not a corporation achieve *synergy*, the enhanced output of people supporting each other in order to produce something greater than the sum of its individual members?

With the right leadership, could not a corporation become something more like a *team*?

These were some of the thoughts that played through Brent Scott's mind as he began his tenure as Syncrude's president in the early 1970s. It took time, hard work, and interplay with other minds for these ideas to cohere, but when they did—and when they were put into action—they powered a Syncrude workplace transformation that was to become a lighthouse to business everywhere.

Unlikely Pioneer

Brent Scott certainly looked and talked like a CEO. He was tall, handsome, positive, and capable of saying a lot with a few well-chosen words. Still, when he joined Syncrude as senior vice president to Frank Spragins in 1972, there was nothing on his CV that suggest he'd be an innovative disrupter who'd lead the creation of one of Canada's first truly transformational corporate workplaces.

Left, Brent Scott with Pierre Trudeau. (Photo courtesy of Syncrude.) *Right*, Brent Scott at his desk. (Photo courtesy of the Scott family.)

Scott was born in Calgary and trained as a civil engineer, graduating from the University of Alberta in 1948. He joined Gulf Canada, working on upgrading the company's refineries in Moose Jaw and Edmonton. He must've inspired confidence in his abilities because in the mid-1960s, he was named senior project engineer responsible for building Gulf's deep-water port and refinery at Point Tupper, Nova Scotia. His experience on that job had a major influence on his later thinking about management.

Scott was initially named senior vice president to Spragins in part because Spragins had no experience building a large organization. By the time Scott became president in 1975, at age 49, he'd effectively been the company's chief operating officer for two years.[19]

He'd never led a large or even medium-sized corporation and came to the job with little advanced training in management. What he did have was a well-disciplined mind, plenty of intellectual curiosity, and, most importantly, the judgment to know a good idea when he saw one. He was a shrewd judge of people, which was vital in his dealings with the Syncrude owners' management committee, and he knew how to spot and seize moments of opportunity to put his stamp on the organization.

The Syncrude organization that Scott joined in 1972 was small and only beginning to grow. It had no vision, values statement, or strategic plan —just a few policies. It hadn't defined the kind of workplace it intended to be. Its culture was effectively a blank slate.

It had two vice presidents (Chuck Collyer in Projects, and Neil Lund in Operations) and a corporal's guard of staff (a two-person human resources department, a few accountants, a single public affairs person [me], and, if memory serves, a single lawyer). Contrast this with, say, Imperial Oil, which had been in business in Canada since 1880 and was a supremely mature corporation with a well-defined culture, procedures, and traditions. I vividly remember my first visit to Imperial's then head office on Bloor Street in Toronto in 1974 The carpets were thick and grey; the lighting was subdued and indirect. There was a hushed reverence about the place, like the anteroom to a temple. People spoke . . . very . . . quietly. By contrast, the Syncrude offices vibrated with energy and laughter.

This lack of structure or traditions presented Scott and his team with an intimidatingly large opportunity to be creative and dynamic—or to fall flat on their faces.

Writing on leadership, US Army General Hal Moore said, "Good leaders don't wait for official permission to try out a new idea. In any organization if you go looking for permission you will inevitably find the one person who thinks his job is to say 'no!' It's easier to get forgiveness than permission."[20]

Or, as Scott was fond of saying, "If you want to get no for an answer, just ask the question."

Moore wrote: "Be dead honest and totally candid with those above and below you." Like the time Scott sat quietly through a long meeting in which the owners offered him trite advice on how to solve a certain problem. He smiled through gritted teeth, then said evenly, "We don't need your *advice*. We need your *help*." In other words, send us more qualified people from your organizations; otherwise, lose the "help."

Scott wasn't one of those executives who love the sound of their own voice and talk too much. He knew how to listen and how to pick his spots to weigh in on a conversation. He had a temper but kept it well sheathed unless it was needed.

When he did bring it out, it was devastating. I clearly remember a 1973 meeting with Scott and Spragins, when extremely tense royalty negotiations were going on across town with the Alberta government. The owners—who were doing the actual negotiations—weren't telling us anything about their progress. Spragins was reluctant to call them. Finally, Scott exploded, literally pounding the desk so hard the coffee cups fell over and several people looked like they were going to faint. "Frank, you're going to fuck it up!" he shouted. "Get on that phone and call them now!"

Scott lived one of General Moore's favorite principles: "Senior leaders who push the power and the decision-making authority down free up talent in their subordinates even as they free up more of their own time to plan ahead." That was one of the central principles of team management, which eventually shaped every level of the Syncrude workplace.

Scott didn't instantly understand what Syncrude needed. He worked his way to understanding piecemeal over the course of addressing several different challenges. The first of these was the recruitment of the thousands of skilled and productive people. In a later speech, he listed some of the challenges of building and operating the industry's biggest megaproject:

The crucial factor, which can't be

legislated or found in a laboratory, is people. Perhaps more than some other industries, we must attract and retain large numbers of skilled and highly motivated people, or we will fail. . . . How will we attract and retain the people to design, build, and operate [this project]? Despite today's unemployment, the people we'll need are in demand elsewhere. We'll have to offer attractive working and living conditions in the sense of remuneration but also job and life satisfaction. Our profits will be modest; we won't be able to afford low productivity or high turnover, so productive and satisfied people are the key to our future." [21]

Scott's vision incorporated both inspirational and negative dimensions. On the inspirational side, he saw that unless Syncrude created an unusual, even unprecedented, kind of workplace —one that would attract and retain good people because it inspired their best efforts and liberated their best energies—the company would fall short.

If that happened, if it slipped into the ultra-conservatized mindset that characterized too many managements of the day, it would become unionized and would ossify into the kind of workplace he'd seen before. In his last job before Syncrude, building the refinery and port at Point Tupper, the entire construction project was wracked with years of wildcat strikes and illegal shutdowns. It was a union job, and in the Maritimes, union members' attitudes were often "if management wants it, the union must oppose it."

The apotheosis of the Point Tupper project was the day not a single unionized tradesperson showed up for work. Management didn't know why, until someone spotted an empty cardboard beer box that had been placed outside the plant gate. On the side of it someone had written "strike today." So, of course, mindlessly, not a single tradesperson showed up to work.

Scott was deeply resolved to prevent that kind of attitude ever taking root at Syncrude. "Management and employees—all of us—need to be together," he told me. "Like a flying wedge." The flying wedge was an ancient football formation popular in the 1920s. It required the offensive team to form a pointed phalanx around the quarterback before the entire formation drove down the field like a flying wedge. It was eventually outlawed because of its tendency to injure too many defenders.

In the beginning, at least, Scott had no idea how to bring that kind of united effort into effect.

The Good Foundation

The first step would be to build a healthy employee relations environment. To do that, he recruited Don W. Scott (no relation), a veteran HR executive from Imperial Oil's IOCO refinery in Vancouver, British Columbia.

Don Scott was a raffish, confident personnel guy who relished the opportunity to build a leading-edge HR team from the ground up. The first two members of his senior team were Norman Cottee, a thoughtful, shrewd benefits and compensation guy, and Dave Simmonds, a lanky, hyper-competitive recruitment manager. Both were in their late thirties, and both had just stepped into the biggest jobs they'd ever held.[22]

And they relished the opportunity. Simmonds was keen to spearhead the biggest high-pressure recruitment campaign any Canadian employer had managed in peacetime. Cottee, meanwhile, was eager to develop an industry-leading compensation and benefits package on a clean sheet of paper.

For Cottee and Simmonds—indeed, for everyone at Syncrude—the immediate challenge was, as Cottee put it, making Syncrude an attractive place to work and Fort McMurray an attractive enough place to live for 3,000 professionals and their families.

For potential employees, Syncrude was an unknown quantity and Fort McMurray was . . . well, let's just say it wasn't an attractant. In those days, most Canadians believed that any community whose name began with "Fort" was probably very remote, very poorly serviced and ... very, very cold.

Syncrude and Fort McMurray needed to attract highly skilled, in-demand people from across Canada and abroad. People from a wide variety of industries. Almost everyone Syncrude needed to recruit had to be found somewhere else.

Don Scott and his team set about their task. Within 18 months, they developed and gained owner approvals for a raft of highly creative and leading-edge progressive employee policies for housing, salaries, benefits, pensions, savings, higher education (tuition and travel support for college-bound children of employees), employee health, and safety. Syncrude's safety program received a major overhaul after the tragic death of two contractors' employees in a construction accident. The company eventually pled guilty to charges of criminal negligence in the case. Brett Scott made radically improved safety practices his personal goal, announcing that "safety begins at my desk."

Thirty years later, a journalist touring the oil sands remarked on the ubiquity of safety-first practices throughout the industry.[23]

The company committed to keeping salaries in the top quartile of a well-paying survey group, and Fort McMurray staff were paid an additional amount called "Zone Differential" to compensate for living costs in a remote location. Compensation practices were fine-tuned to recognize the value of individual progression and performance.

And then there was housing. Syncrude would be moving thousands of people to Fort McMurray. Hardly any new housing had been built to accommodate them, and some of the temporary housing that was available was third rate. "One project engineer, whose trailer twisted on its foundation, resulting in a bedroom door that would not close and a bathroom door that would not open, finally gave up and left," Cottee reported.

Despite the owners' initial reservations, Syncrude brought in interest-free home purchase financing for employees and created a subsidiary, Northward Development Ltd., to quickly build large amounts of employee housing. John Elson had been a developer, and he had a developer's energy and drive. He was made Housing VP.

A foundation of good HR policies and practices would be *necessary* for the creation of a flying wedge culture, but they wouldn't be *sufficient*. Something more was needed. The hunt for that something took Don Scott and his team to the field of industrial psychology.

It was the 1970s. Industrial psychologists and other social scientists were beginning to gain wide public attention in universities and in business. Human resources people, and even enlightened executives, were reading people like Abraham Maslow and Peter Drucker and writers like the social psychologist Douglas McGregor, author of *The Human Side of Enterprise,* which, as Cottee put it, "provided a simple framework with which managers could examine the assumptions they held about their employees' motivation and adapt their environment."

The table was about to be set for the creation of workplaces designed from social science-driven research.

Herb and Jonno

In the early 1970s, most Canadian corporations, particularly in the oil and mining sectors, were traditionally managed, which is to say they were managed by executives who believed in hierarchical management in which the watchwords—usually not spelled out but definitely enforced—were "I decide. You follow." In management studies, this became known as Theory X Management.

Just as the doctrine of the Divine Right of Kings inevitably led to revolutions and the development of democratic institutions, Theory X Management produced push-back in the form of the organized labour movement.

In many industries, this led to a highly adversarial workplace environments in which managers fought to preserve management rights and unions fought to preserve union rights. What often got lost in this struggle was the best interests of the enterprise. Sometimes workplaces became toxic and both sides lost sight of the need for rational policies. Militant unions and obdurate employers perpetuated this.

In the 19th and early 20th centuries, the mining industry was dominated by two-fisted managers and equally two-fisted unions. Managers did little to earn employee trust, so employees looked to unions to defend their rights and interests. John D. Rockefeller, for many years the most hated businessman in America, got that way by sponsoring dictatorial mine managers that, during the Ludlow Massacre, employed federal troops and Pinkerton guards to shoot down rebellious union workers and their families. Afterward, the Rockefeller Foundation studied how to improve industrial relations. One of its studies, *Industry and Humanity: A Study in the Principles Underlying Industrial Reconstruction*, was written by William Lyon McKenzie King, future prime minister of Canada. King didn't like unions. Instead, he advocated improved industrial relations based on employer paternalism and conciliation.

Enter Brent Scott and what he learned in the course of building the refinery and port at Point Tupper. Point Tupper was a union job and a toxic workplace par excellence, a place where union leaders had lost sight of the needs of the enterprise and made a habit of opposing management purely for the sake of it.

So it was that Brent Scott asked Don Scott to look for a consultant and advisor who could help Syncrude avoid this trap and grow into the kind of thoroughly modern, rational, consultative organization he envisioned. And that was when Don Scott first heard the name Herb Shepard.

Herb Shepard at home (circa 1976). (Photo courtesy of Jonno Hanafin.)

Shepard, a behavioural scientist from Hamilton, Ontario, was one of the founders of the budding field of organizational development. He graduated from the University of Toronto before receiving his PhD from MIT's Sloan School of Management. One of his doctoral advisors was Douglas McGregor (see above), who happened to be a student of Abraham Maslow.

In the early 1950s, Shepard was hired by Standard Oil of New Jersey (now ExxonMobil) as a personnel researcher. This was the first ever internal organizational development (OD) consultant position. Shepard took part in pioneering social sciences work at Esso refineries in the US Gulf Coast. He and other behavioural science luminaries of the time created the T-Group (the Training Group or sensitivity group) as a revolutionary methodology for personal growth and development.

Shepard's work at ExxonMobil and his subsequent work at California's TRW Corporation, one of the first high-tech companies in the aerospace sector, laid the foundation for many concepts in the field of OD, including the Team Concept he brought to Syncrude.

Don Scott was an old Imperial guy, and doubtless his contacts there had contacts at ExxonMobil, which is how Shepard was invited to Edmonton. He was a masterly consultant and made a very positive impression on Brent Scott, Don Scott, and the rest of the Syncrude executive team. He began to meet with them every three months to explore how Syncrude could create a team-based culture.

There was no blueprint for doing this. Given the scope of the culture work envisioned by Syncrude executives, Shepard suggested that management add an OD manager to its HR team, a person who could, with Shepard's consulting guidance, spearhead the huge effort to scope out and implement a team-based organizational culture.

The initial cross-Canada search for such an OD specialist came up dry. There were only a handful of professionals in the budding field. Most were based in Eastern Canada, and none were interested in relocating their families to Edmonton or Fort McMurray, nor did they want to work on a high-risk project. Don Scott asked Shepard to help him find the right person.

Shepard had always made it a point to stay involved in university teaching and had taught at several US institutions; in the course of his teaching, he always had an eye out for people in whom he could discern some sort of spark. He'd seen such a spark in a young IBM systems analyst named Jonno Hanafin, who was doing his graduate studies. In 1974, Shepard brought him to Edmonton to meet with the Syncrude executive.

Hanafin as he then was. (Photo courtesy of Syncrude.)

Shepard knew a thing or two about how oil industry people thought. He understood Don Scott and Brent Scott and what they needed. In conversation, his manner was very information-seeking; he knew when to listen—to really listen—and when to speak, and when he did venture an opinion or a recommendation, it was always in a way that was respectful, pleasant, and well tailored to his audience.

Hanafin, his nominee, was slight, quiet, and gently plainspoken. And at 25 years of age, *very young*.[24]

A Look Back with Hanafin
How did you feel starting the job?
It was an incredible expression of trust in

Shepard for Syncrude to offer the position to me. I had no formal OD experience, only a few years of apprenticing with Shepard, one intense summer immersed in [the National Training Laboratory's] consultant development curriculum, and—in Shepard's judgement—potential and great

Hanafin today (2020). (Photo courtesy of Jonno Hanafin.)

instincts. I was eager to make a right-angle career change, so I quit IBM and headed to Edmonton with the understanding that Shepard would continue on as my coach and guide.

What was your understanding of Team Concept at that time?

Syncrude had a tonne of big challenges. Technologies that had never been tested at that scale. A hostile climate. A very complex set of systems. A new workforce with no shared history. The prospect of a massive start-up to get the whole thing going. And the mining industry's history of bad labour relations.

What did Herb Shepard base his theory of team management on, exactly?

Herb was a social scientist, and he believed that social science research supported certain assumptions of how team management could work. He taught that all teams are groups; all groups are not teams. When developed intentionally, teams perform better than the sum of their individual members. He had a systems perspective: everything is connected and impacts everything else. This means people should look at what's best for the whole, not each individual part of a unit. In other words, we need to get away from tending to one's own territory.

Did he have some kind of general theory of organization effectiveness?

Sure. He thought organizations contain all the information they need to identify obstacles and resolve problems effecting organization performance. He thought the primary reasons this information isn't used is because managements lack attention, skill, and leadership maturity. He thought the best way for organizations to capitalize on all their people resources was to create social experiments that tapped everyone's actual experience at work and to invite people to participate in improving everyone's performance and satisfaction.

Where did employees' rights fit into all this?

If employees are treated with respect by management and feel their voices are heard and considered, they'll feel less need for an outside agent to advocate for them. Employees will do it directly and without paying dues to a third-party representative.

So apart from reducing conflict between management and employees, how did you think team management would benefit the organization?

A team-based culture would place a premium on win-win collaboration rather than win-lose competitiveness among departments, divisions, or, for that matter, individuals (or employee and supervisor). With the right skills, commitment, and support, co-operation between groups would result in additive synergy instead of destructive competition. Also, the high degree of risk and complexity in Syncrude's project meant we needed as many minds as possible focused on problem identification and problem-solving. Neither was "management's job"; both were *everyone's* job. Shepard taught that the more supervisors, managers, and leaders who were involved in decision-making, the better the decisions that would result and the more committed everyone would be to executing the decisions.

My sense is that many of those ideas are generally accepted today.

True. But in the 1970s. they were seen as cutting-edge, if not revolutionary.

So how did you and Shepard go about making all this happen?

It took more than five years of intensive work with hundreds of people. Since there weren't any best practices or precedents to follow, it was largely "learn as we go." We took what behavioural scientists call an "action research approach." Assess the situation, offer suggested improvements, try them out, notice the impact on the group and the process, track learning, and rinse and repeat.

Where did you start?

Without a proven playbook, it was total improvisation. The challenge for leadership and HR paralleled nearly every other function in the Syncrude start-up. Syncrude was an organization populated by individuals way over their heads. There were project experience and engineering drawings and proud statements of vision and intent, but the fact was no one had ever done anything quite like this before. This dynamic created a bonding experience of "we are all in this together." From an applied behavioral science perspective, there was no predetermined place to start, so you could start anywhere, or so it seemed.

HR had put good policies and programs in place, so we decided to start with developing criteria for performance appraisals. What kind of behaviour was expected of Syncrude leaders in the new company culture? This was one of the first opportunities for establishing cultural specifications. People do what they get rewarded for, so here was a chance to begin shaping leadership behaviour. As they say in the culture-building business, birth is easier than resurrection.

Where did Brent Scott and management fit into this?

This was Scott's master stroke. With the encouragement of Shepard and the support of his colleagues, Scott recognized the opportunity and peril inherent in bringing two industries (oil and mining) together on this scale for the first time. If an organization's culture is "the way we do things here," Scott was determined not to spend endless time arguing over which industry's way of doing things was better. His charge to HR generally and me specifically was to create a company culture that captured the best of the two industries' best practices, plus a third element that was designed for Syncrude's unique needs and possibilities.

It was tempting to create a reward system based on appraisals at headquarters, but I hesitated. I was new and unproven. The only credibility I had was Shepard's strong advocacy for me. In one of my first consultations with Shepard early on, he agreed with me: I needed to have a small success to start. But what? A second bit of wisdom from [him] was "load experiments for success." So that's where I started. I wanted a ready-made project out of the limelight where I could do that.

I don't remember hearing about this at the time. Where did you go from there?

Ron Goforth [director of research and environmental affairs] wanted me to do some team building with his senior people. These were scientists interested in learning how to work together better, a more appealing idea than officiating trench warfare. The team development session was a rousing success.

Goforth spread the word, and a model was created for other functions to experience early tastes of the Team Concept. Shepard facilitated a similar team development for the HR leadership team. Then we took on mining and environmental affairs, who'd been at loggerheads for some time about how to clear the surface of the mine area to minimize the impact on residential wildlife. The conflict was intense and lingering. Looking back on it was hardly "loaded for success," so maybe it wasn't the best choice for a first test. But it was a modest success and didn't damage my reputation too much.

So designing the culture started at the top. How did you first work with the leaders?

Our way in was the development of the performance appraisal process. Performance appraisal can be a powerful tool for encouraging certain behaviours and rewarding people for contributing to the team-based culture we were trying to build. So we interviewed all Syncrude senior executives about their views on that. We played back the interview data to the leadership group for decision-making. They then created their own template for how they would evaluate the contribution of all Syncrude managers and leaders to supporting the Team Concept and values. This constituted an early step in birthing the team culture.

Herb Shepard and I kept facilitating the development of Brent Scott's senior leadership team. At this point, the bulk of Syncrude's work involved plant construction on site in Fort McMurray and prefabrication in Edmonton. I partnered with Bechtel's OD team from San Francisco to design and facilitate inter-group initiatives between the Syncrude project team, led by Chuck Collyer, and his Bechtel counterparts. These regular interventions enhanced an already well functioning relationship between the two organizations.

So that was the leadership group. How and when did you start to push Team Concept down through the whole organization?

One item in Herb Shepard's strategic guidebook was "innovation requires a good idea and a few friends." I eventually did this formally by building a staff, but first, I wanted to create an informal support group from throughout the Syncrude organization.

I got the idea of how to do this from a story Shepard told of how he had built a PhD program he founded in Cleveland, Ohio. Since it was a new program with the usual growing pains, Shepard put together a cross section of students, teachers, and administrators to solve problems. They came to be known as the Hats Group. When they assembled to work issues, everyone took off the hat of their role or experience or function and put on the hat of the program; what will work best for the overall program? They were remarkably successful in navigating the start-up of the PhD program, which today is recognized as one of the foremost offerings in the world.

So over the course of his first six months at Syncrude, I made mental notes of people who stood out by virtue of their sparkle, for lack of a better word. I invited about 20 of them to join an optional and informal group dedicated to supporting and catalyzing the effort to build the Team Concept culture. We convened the group for a weekend team-building workshop at the Red Deer Inn hotel. Shepard and I co-facilitated.

Take me back to that moment and how it worked.

One of team's tasks was to construct a bridge over the hotel's swimming pool using only arts and crafts materials. It was poetic to see John Lynn, Syncrude's project site manager—the man responsible for building the new bridge over the Athabasca River to transport heavy vessels—contributing leadership on the pool task. At the end of the weekend, everyone had the option of signing on as a co-conspirator of change or declining and wishing the team well. All but one of those attending signed up for the mission and joined what became known simply as Syncrude's own Hats Group.

It sounds like a cross between a club and a secret society.

[Laughs] We had no secret handshake. Just a greater purpose than their specific job. You might say we saw ourselves as part of a positive conspiracy' in the sense of the Latin verb *conspirare*, "to breathe together." Members of the group met in subgroups and occasionally as a whole when requested by one of the group's members. We often arranged to help their homework groups solve problems and resolve conflicts. It's true we did this under the radar but with senior management's support. Brent Scott knew of our existence and came to one of our early sessions to acknowledge the members unofficially, thank them, and wish them well. His visit and encouragement were inspiring. The Hats Group lasted beyond the start-up of the plant.

Hanafin, with John Howard (standing). (Photo courtesy of Jonno Hanafin.)

In time you headed a division of HR called Organizational Development. How was that built?

In 1977, there weren't many OD professionals available. I started by bringing in Michael Kitson, a grad school friend and fellow Shepard protege from DuPont. I then followed the same model Shepard and me used to hire trainers and others with potential, pairing each of them with an experienced external consultant. Each of the consultants were given the charge of coaching their internal partner. Each pair was assigned to a unit of the operation: Upgrading and Utilities, Mining, Projects, Administration. The HR group's official name was Management and Organization Development, but we referred to ourselves as the Mod Squad after a popular TV show at the time. We were a team of eager early career adventurers sharing a gondola going up a steep learning curve.

I guess there was a potential for the Mod Squad to be seen as a little precious or at least strange.

The Mod Squad. (Photo courtesy of Jonno Hanafin.)

As identified change agents, we enjoyed a camaraderie of difference. We had to be a bit different, if only to model what we were advocating. The trick was not to be *too* different. If we were too different, people would focus on how different we were, not on what we were proposing. We developed a code to signal each other if we were being too different. I coined the phrase "Perceived Weirdness Index" (as in, "Your PWI is too high"). The phrase originated at 2:00

a.m. one evening at an off-site program when one the Mod Squad women drank a table full of mining supervisors under the table. Impressive, but not the kind of credibility they wanted to create.

The team experimented on itself with new processes as a way of testing and creating teamwork theory. Since I was manager of M&OD, I was responsible for completing performance appraisals and recommending salary treatment; we got the idea of doing both of these activities in and by the group as part of an annual team development session. Hopefully, the statute of limitations has expired.

Mod Squad members were based in Edmonton and spent a large portion of our time in Fort McMurray working with their client leaders. We had a Northward Developments' apartment in Fort McMurray, so team members overnighting could eat, compare notes, and socialize together. We bought an old truck with lots of character for driving back and forth to the plant site. We called it The Grey Ghost.

In addition to supporting and teaching their internal consulting partners, the external Mod Squad members were given client assignments in the Fort McMurray community. One consulted with the hospital board, one with the school board and other expanding community organizations. Coincidentally, one of the members of the Hats Group who lived in Fort

McMurray, Grant Howell, was elected to town council. Howell was a crucial link pin among the company, the plant, the community, and local government.

Are you saying this changed the character of Fort McMurray?

I wouldn't go that far, but I believe we had an impact on the community. When the 2016 wildfires almost destroyed the city of Fort McMurray and 80,000 people had to evacuate, a lot of people commented on everyone's tremendous teamwork and caring for the community as a whole. So maybe we modified the community's genes a little.

So tell me how you went about driving team management down to the everyday workplace.

It had two parts. A five-day off-site leadership training workshop for the 600-person leadership group—not all of them at once, of course—and then a two-day Team Skills programs to every employee, all 4,000 of them.

Why did you start with the leadership?

Culture is leadership at scale. It would be up to every VP, director, manager, and supervisor to live the Team Concept. When a leader's words and music do not match, people go with the music. All the philosophy and vision statements represent the cultural design specifications, the aspiration. It is the leaders' day-to-day behaviour that constructs the actual edifice. We had to equip everyone in a supervisory capacity with a felt understanding of what Team Concept was. How leaders were expected to function within the new culture. The program was called Managing in Sync (MIS.) Everyone in a management or supervisory role in the company would be signed up for the program. It would take place off-site in the Rockies near Jasper.

Getting management support for an intervention this big and expensive must have been an adventure.

John Howard, VP human resources who took over after Don Scott left, was responsible for the design and delivery of the experience. We took the proposal to Doug Hingley, the senior VP for administration, for approval. Hingley was an accountant by trade and charged by the owners with keeping an eye on the growing Syncrude budget. We knew it was going to be a tough sell.

Howard warned me that Hingley thought five-day workshops were too costly and should be reduced to three. When we met, he said "I have a real concern about the length of the program." I said, "You're right, Doug. This agenda may be overly optimistic and by all rights should be seven days, but we are reluctant to take people away from their families for that long." I remember Howard freezing like a deer in the headlights and held my breath as Hingley thought. After what seemed like an hour, Hingley finally said, 'OK, I just wanted to check. Go ahead.'

When we left the meeting Howard just shook his head and said, "I don't believe you did that and got away with it." [Laughs]. A bit of politics for a good cause. I guess I should say to people, don't try this at home.

Can you tell me a bit about the MIS program?

Every leader in Syncrude, president to first line supervisor, was invited by Brent Scott to attend a session of the Managing in Sync program. Each session included 32 participants arranged into four learning teams. The program went from Sunday to Friday and was held at the Overlander Inn just outside the gate to Jasper National Park.

The Overlander Inn. (Photo courtesy of the Overlander Inn.)

On Sunday morning, the participants from Fort McMurray boarded a vintage DC-3 aircraft that took them to Edmonton. There they joined the Edmonton-based participants at the Edmonton Plaza Hotel. Brunch was served, and Brent Scott appeared every week to welcome the group to the program, share his vision of the Team Concept, and encourage them to be open to learning, something he expected of all Syncrude management. After brief remarks, Scott took a few questions from the group (most of whom were meeting the president for the first time).

I remember one priceless comment to Brent

Scott by a first line supervisor in mining. He said, "Thank you, sir, for coming to be with us and for this opportunity you are giving us. This is the first time I have heard you speak. I believe you mean what you say. And I want you to know I am with you. But I have to tell you there must be some SOB between me and you that does not agree because this is the first time I am hearing what you just said."

The program began with an open bar reception and dinner. One of the few regrets of the MIS program was that we did not photograph participants faces when they first heard this was an open bar. This program feature immediately lowered resistance in the group and started building trust with the faculty. The program had morning, afternoon, and evening sessions. It went from 8:00 a.m. to 10:30 p.m. every day. They could drink free, but they had to work for it. Sunday evening included introducing the teams. The group consisted of VPs, directors, managers, and first line supervisors. They had been divided into four learning teams by shared identity, upgrading, mining, administration, and other. There were eight people, plus an external OD consultant, in each group.

There were five advertised objectives for the MIS program. We said, "You will leave with (1) [an] understanding of what a team is, how it works, how to manage it; (2) skills in team membership, leadership, [and] problem solving;

(3) [the] opportunity to look at your own management style; (4) an action plan for back home work; and (5) memories of an enjoyable learning experience with your co-workers."

1977: The first MIS workshop. Shepard's on the bottom row to the left, and Hanafin's holding the flag. (Photo courtesy of Jonno Hanafin.)

Staying with Shepard's counsel to load experiments for success, the first MIS class in January 1977 was proposed as a pilot. The participants were hand-picked as, if you will excuse the abomination of metaphors, the cream of the canary in the mineshaft crop. They came with a critical eye, an open mind, and a supportive heart. They were, as were all who would follow them, instrumental in fine tuning the MIS program as it rolled out. Being in their home group was supportive and built [off] their sub-culture. Going out from their home group was a chance to engage their curiosity. How much of what they have been hearing about the others was true? And what is this larger Syncrude thing we share?

The design of the week included topics matching lively issues in the organization as it took shape. It was a highly experiential workshop. Over the course of the week, there were eight tasks to be done, each with a designated group leader. The leadership rotated and every participant had the chance to lead their group. Each stint as leader included feedback from the group on how they led.

The basic design flow was as follows:

MIS Workshop Design	
Sunday	Orientation to Each Other and the MIS program Preliminary Synergy Exercise Forming a Total Community
Monday	Why Team Concept? Arctic Team Survival Exercise – Team Synergy Team Project
Tuesday	Transactional D|Analysis – Human Behaviour Communication Practice Application in the Organization Personality Profiles
Wednesday	Creative Problem-Solving Assimilating New Team Members
Thursday	Managing Commonalities and Differences Systems Thinking Back Home Planning High-Performing Systems
Friday	Dealing with Stress Commitment—Letter to Self Team Advice to Self—What are We Going to Do Now?! Group Photo and Departure

We planned several activities during the week:

- concept presentation followed by small group discussion
- group assignments and presentations to all
- giving and receiving feedback from one another and staff
- engineering and building a tower, in project roles, made from craft supplies
- watching movies to spark metaphors and

> learn about change

- an arctic survival exercise where teams were stranded in Northern Canada and had to figure out how to survive (with specific scores measuring team work)
- learning creative problem-solving techniques, then applying those methods to solve real time problems (like improving family morale, dealing with the bears and crows at the Fort McMurray dump, building *esprit de corps* in the utility plant)

The MIS program in the mountains was all about knowing each other, understanding what a team-based culture meant, and learning together. Working in a new and stressful place was balanced by a shared positive experience. It created bonds that lasted for years. And these bonds and insights served as the grout in joining together diverse trades, origins, cultures, and individuals into a successful Team Concept culture. The team culture contributed to outstanding effort, commitment, and performance over the project.

By the time the MIS program was driven through the whole Syncrude organization, more than 600 managers and leaders completed the program and received a diploma from Brent Scott that said they had "made a significant contribution to the initial stages of the development of Team Concept in Syncrude." Indeed, they had.

A good story. Did Team Concept take at Syncrude? Did it become central to Syncrude's culture?

The best evidence for yes is that that brand-new, untested organization built the Syncrude plant on time and on budget in 1978 and operated it successfully. A remarkable feat in any era.

On the day I left Syncrude in May 1980, John Howard gave me a sealed going away present and made me promise not to open it until I was on the plane. It contained the results of the employee vote on the very first unionization drive at the plant. Employees resoundingly rejected having someone speak to management on their behalf. They felt involved, heard, and respected by management. They didn't need a translator.

In the following years, by secret ballot, Syncrude employees turned down five more union-organizing attempts. That was their answer to whether team management met their needs.

An Editorial Footnote

The power and lasting quality of Syncrude's team culture was born out many times in the years to come. In the next chapter, you'll see how subsequent CEOs built on teamwork to progressively streamline and modernize the company, cutting costs and improving productivity without resorting to the kind of brutal top-down cost cutting and organizational flattening that came to characterize the worst of American-style management in the last 40 years.

Was Syncrude's transformational workplace culture taken up by the rest of the oil sands industry, the oil industry in general, and the world at large? I don't know.

Did it influence better-known and more recent workplace culture innovations in Silicon Valley and North American industry in general? Perhaps it did. It's said that the flapping of a butterfly's wings in Ecuador can influence the birth of a hurricane in Asia.

Looking back, Hanafin remembers Syncrude as a unique life experience:

There is something special about a community of strangers that come together for a noble purpose and a limited time, bond over a common challenge, give the effort their all, celebrate (hopefully), and then move on. For me, the challenge, the adventure, the shared success and mostly the people of Syncrude have been a tough act to follow. They continue to motivate me to try something new and make things better. And when you can do that in conjunction with people who come to care about one another, it does not get any better.

9. THE GATHERING STORM

Early Opposition

Prior to the late 1960s, there was opposition to oil sands development, but it had nothing to do with the environment. It emanated from within the industry itself and was driven by a mercantile interest. Conventional oil producers —companies that drilled for oil the old-fashioned way—worried that a sudden upsurge in synthetic crude from the oil sands would drive them out of the market because the Alberta government would require them to cut back conventional production.

The so-called independents, to distinguish them from the multinationals or majors, saw the oil sands as a potential competitive threat. So when the regulatory agency that had to approve applications for oil sands projects held hearings, the independents vigorously intervened to prevent or delay government approvals of oil sands permit applications. They were successful from 1959, when they convinced the Oil and Gas Conservation Board to defer Syncrude's application in favour of GCOS, to 1968, when the board finally gave a conditional green light to a later Syncrude application.

The next opposition was political and came from the nationalist left.

The Waffle and the National Energy Program

The idea that multinational oil companies were inimical to the interests of Canada first became politically important during the 1970s. The prophets of this point of view were nationalistic economists like James Laxer and Mel Watkins of the University of Toronto, and their ideas were taken up by both the so-called Waffle Caucus of the New Democratic Party (NDP) and the left wing of the Liberal Party of Canada. The most important of these Liberals were Pierre Elliott Trudeau and Marc Lalonde, Trudeau's Quebec lieutenant in the Liberal cabinet. The same theme was echoed in the media by Mel Hurtig, a Liberal gadfly who was accomplished at self-promotion but not at getting elected. Politically, the high-water mark of this school's influence was the Trudeau Liberals' creation of the so-called National Energy Program, which ran from 1980 to 1985.

The leader of the NDP, David Lewis, organized

a series of cross-Canada hearings on the subject of what the NDP called "corporate welfare bums," a category to which Lewis assigned the major oil companies. The so-called hearings, roughly modelled on Bertrand Russell's and Jean-Paul Sartre's 1960s notion of an international war crimes tribunal, was a travelling propaganda platform for Lewis to hector any large and foreign-owned)corporations prominent in the local community. I believed Syncrude would be on Lewis's hit list anyway, and I advised Spragins to appear before Lewis's hearing in Edmonton, where he was allowed to speak briefly and was then made to endure several rants about the sins of multinational corporate welfare bums before the panelists packed up their papers and caught the plane to the next city and their next auto-da-fé.

The Waffle Caucus and the Lewis hearings left a lasting impression on the left wing of the federal Liberal Party. When Prime Minister Trudeau unveiled his National Energy Program in 1980, its stated goals were to shelter Canadians from the vicissitudes of international oil prices and to provide some limited opportunity for some Canadian oil suppliers (in the oil sands) to have access to higher oil prices. By the later admission of Lalonde, his senior lieutenant, the unstated goal of the National Energy Program was to confiscate a major part of the Alberta government's oil revenues and divert them into federal coffers.[25]

The National Energy Program touched a responsive chord with Liberal and centre-left voters, especially in central Canada, and it had the effect of shutting down oil sands development, after Syncrude, for a good five years. It also triggered a lengthy and at times vicious interregional debate in Canada that eventually led to a tectonic shift in federal politics that included the rise of the Reform Party of Canada and a realignment of the federal Progressive Conservative Party. One legacy of this shift was the enduring suspicion among many western voters that the Liberal Party was the prisoner of central Canadian elites and not to be trusted, ever.

The First Wave of Environmental Criticism

In the 1960s, the first wave of environmental concern swept through the Western world. "Environmentalism" and "ecology" became the watchwords of educated people. The interrelationship of economic activity, environmental quality, and human health began to influence human consciousness. In 1970, the Nixon administration created the Environmental Protection Agency. In 1971, Greenpeace was founded. In 1972, the year Syncrude received its first go-ahead, the Club of Rome published the first sustainability manifesto, *The Limits to Growth*, and the

United Nations Environmental Program was created. By the time Syncrude got its first go-ahead, environmental concern was in the air everywhere, and it was an increasing preoccupation of politicians and businesses.

Environment was news, from media pictures of strip-mine moonscapes in West Virginia to magazine spreads showing the flaming pollutants on the surface of Pennsylvania's Cuyahoga River.. Everyone was told, and everyone believed, that traditional practices of industry were disastrous and needed to seriously change. This change of consciousness was to have very far-reaching consequences for business, government, and consumers around the world.

All of this consciousness-raising was on the rise when Syncrude finally got its go-ahead from the Alberta government and Syncrude's people began the serious work of designing the complex.

The First Defender

Syncrude's response to this first wave of environmental consciousness was led by Ron R. Goforth, the company's first director of research and environmental affairs.

A native of Arkansas, Goforth graduated with a PhD in analytical chemistry from the University of Alberta and joined Syncrude in 1971 at the age of 33. Like many people in

Goforth, circa 2000. (Photo credit: The Goforth Family.)

Syncrude's original team, he was stepping into the biggest job of his life. Goforth was a large, soft, cheerful man of boundless intellect and interests. He was a scientist and took the scientific method with the greatest seriousness. He saw it as his responsibility to read and understand the environmental science literature of the time and have a solid grasp of the ecological implications of the Syncrude plant design. He put together an environmental affairs team that could address these implications. Goforth returned to his native Arkansas in the mid-1980s and died there in 2012 at age 74 after an active career in university teaching, science consulting, and teaching travel in the Pacific.

He viewed the challenge of meeting environmental criticism with his customary zest.

The Original Environmental Issues

Goforth believed that the environmental issues raised by the original Syncrude design were: (1) the high air emissions of sulphur dioxide from the upgrading plant, (2) the impact on the land of the huge Syncrude mine, (3) the hazard that the tailings pond would pose to migratory wildfowl, and (4) the drainage and rehabilitation of the tailings pond. He also believed that Syncrude needed a factual and vigorous communication program to address public concerns about these issues.

Goforth was under no illusion that these challenges could be easily surmounted. The bitumen in the oil sands contained a lot of sulphur—5 to 9 percent by weight, making it some of the highest sulphur-content petroleum in the world. Nobody wanted that amount of sulphur in the final synthetic crude. Some of it could be taken out in the upgrading process. The rest of it had to go *somewhere*, likely into the air; 250 long tonnes of sulphur dioxide per day went up the company's 600-foot stack. What would the downwind consequences of that hot plume be for the forests of the northeast? Would Syncrude become the world's newest major source of acid rain, the hottest environmental issue of the day?

As to the mine, it was going to be an ugly thing: a pit 38 square kilometres in area and 50 to 100 metres deep. The world had seen too many huge, abandoned mining pits, like the old Anaconda Mine pit near Butte, Montana, and no one wanted another one.

The old Anaconda cooper mine at Butte, Montana. (Photo courtesy of the Mining History Association, https://www.mininghistoryassociation.org.)

As for the tailings pond: well, that was going to be an even bigger problem, though no one appreciated by quite how much.[26]

An oil sands extraction plant is a gargantuan consumer of fresh water (once heated, to separate the bitumen from the oil sand) and an equally gargantuan producer of waste water contaminated with tailings (sand, clays, and residual bitumen). Syncrude's engineers designed a water-management system that, on paper, would minimize withdrawals of water from the Athabasca and, once the sand and clays settled on the bottom of the tailings pond, enable the company—many years down the road—to drain the tailings pond and turn it into a wetland or fen.

A key assumption of this design was the settling rate of the clays. The tailings pond was designed for 25 years of mining and extraction. Final reclamation of the pond was years away; the first challenge was the immediate safety and environmental risk of the pond itself.

Safety first. It couldn't leak contaminated water into the region's groundwater or the Athabasca itself. And it had to be physically secure, meaning the dykes holding it back —dykes built of clean sand left over from extraction—couldn't fail.[27]

We will return to subject of the tailings pond in a moment.

Sulphur Emissions and Acid Rain

Goforth had little impact on reducing sulphur emissions.[28] Sulphur emissions (SO_2, actually) were the consequence of the design of the upgrading plant, in particular its use of fluid cokers to crack the bitumen into sour crude oil for later hydrotreating. Fluid coking was a licensed Exxon technology and thought to be the best available for first-stage bitumen upgrading.[29]

Mindful of the need to minimize sulphur emissions, the original plant designers looked at flue gas desulphurization, which was then an emerging technology; they judged it as immature and costly and so passed on it. Syncrude adopted flue gas desulphurization for one of its three upgrading plants in 2008 and subsequently reduced its overall sulphur emissions by 40 tonnes per day.

While Goforth's environmental team was unable to impact the level of sulphur emissions, they did develop an innovation that improved the monitoring of the environmental effects of SO2 downwind of Syncrude in the boreal forests of northern Alberta and Saskatchewan. Lichens uptake sulphur dioxide and can be used to measure how much of it is coming down from an industrial plume. Syncrude was one of the earliest companies to build a widespread network of lichen stations for this monitoring.[30]

In the 1970s, the hottest pollution issue was acid rain. It was most acute in parts of eastern North America where soils were especially vulnerable to acidic fallout from smelters and heavy industry. Goforth was acutely aware of this issue, given the sulphur dioxide emissions of the Syncrude stack and its potential to acidify forest soils downwind. One of the objectives of the lichen-monitoring network was to be an early warning system for soil acidification.

Research eventually showed that these downwind forest soils had significant buffering capacity from acidification and that, in fact, many of the soils likely to receive aerial sulphur deposition were sulphur-deficient from a plant nutrient perspective. Acid rain turned out to be a non-problem for the oil sands.[31] In time, the most authoritative scientific study of the environmental impacts of the oil sands, authored by the Royal Society of Canada (RSC) in 2020, judged acid rain to not be a major problem.[32]

The two challenges to which Goforth and Environmental Affairs made the greatest contribution were reclamation of disturbed areas (including the company's huge mine pits) and reduction of the tailings pond's risk to wildfowl.

Syncrude's work to reclaim and revegetate disturbed areas was outstanding.

Above, reclaimed mine areas. (Photo courtesy of Alex MacLean).
Below, after (Photo courtesy of Syncrude.)

Goforth was on top of the risk that the tailings pond would pose to migratory wildfowl, and very early on, he commissioned one of the continent's leading experts on wild bird behaviour to study how wildfowl migrated through the oil sands region and how they were potentially attracted to water bodies for rest and feeding. The consultants reported that there were two aspects of bird behaviour that made the tailings pond a potential menace. When the birds came through the region in the spring, parts of the tailings pond—kept free of ice by the continuous inflow of warm tailings fluids—would be ice-free and thereby the only open waters available to the birds for rest. And there would be rafts of floating oil on the surface, a death-trap for any bird landing there. The second aspect of bird behaviour that made them vulnerable was the tendency of the flocks to fly at night, increasing the risk that they wouldn't see the floating oil.

On the strength of this research, Goforth's team developed a strategy to minimize the hazards of the tailings pond. Syncrude's engineers were asked for process improvements that would reduce the amount of bitumen escaping into the tailings pond. Pond skimmer machines were created to remove floating bitumen from the pond surface.

Bird deterrent device, a.k.a., "Bitu-man."
(Photo courtesy of Syncrude.)

And a unique kind of floating scarecrow —dubbed "Bitu-man"—was created to frighten the birds away with noise (propane cannons), flashing lights, and a brightly-lit body.[33] Hundreds of these scarecrows were built and placed around the pond.

All of these measures didn't eliminate the risk of the tailings pond, but they ameliorated it. And amelioration, using the best science available at the time, was what Syncrude (led by Goforth) was about. No measure could guarantee that the tailings pond would never endanger a single duck. As events were eventually to prove.

Wildfowl at Risk

The oil sands tailings ponds are located on one of the major continental inland flyways for North American wildfowl.

It's estimated that each spring, 2.6 billion wildfowl fly across Canada, migrating to feeding and nesting grounds in the north.[34] For

27 years, Goforth's deterrent system prevented most of those birds from landing on the open waters of the Syncrude tailings pond. Then, one fateful day in spring 2008, the system failed.

Pacific Flyway. (Photo source: Ducks Unlimited, https://www.ducks.ca/migratory-birds/.)

On that day, a blizzard slowed the deployment of the deterrent devices just as a large number of ducks flew into the area on their way north. Thousands of them tried to land on the pond. At least 1,600 were caught in rafts of floating oil, and they died painfully. Few could be saved.[35]

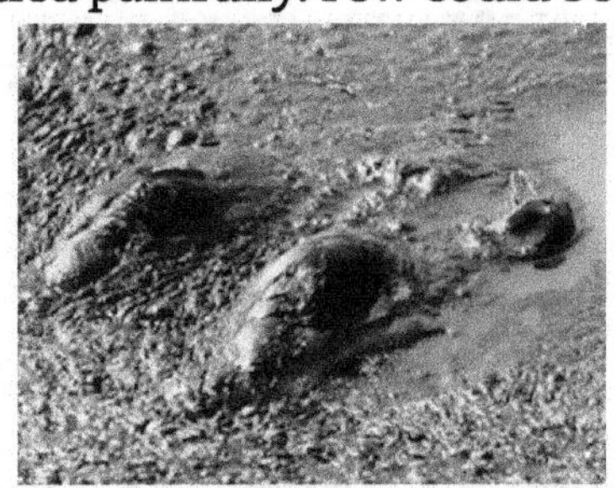

The price of system failure: oiled ducks in 2008. (Photo credit: Todd Powell, Alberta Fish and Wildlife.)

For anyone who likes ducks—and who doesn't?—it was a heartbreaking event. It was also a public relations disaster. Round-the-world headlines and tragic pictures followed, along with justified denunciations of Syncrude and the oil sands industry. Syncrude's tailings pond was only one of several on the flight path. How many other wildfowl disasters went unreported?. Politicians and environmental groups called for punishment, if not an outright moratorium on oil sands development. Eventually, Syncrude was fined $3 million, much of which was donated by government to the University of Alberta for bird deterrence research as well as to conservation organizations and the environmental program at Keyano College in Fort McMurray. According to the federal environment minister of the day, the late Jim Prentice, the fine was the largest in Canadian history for an environmental offence.

Syncrude's and the Alberta government's handling of the oiled ducks issue drew major criticism.[36] Both seemed flat-footed and slow to respond. The response of Alberta's premier Ed Stelmach—he minimized the incident, contrasting it to the millions of ducks killed every year by hunters or in collisions with buildings and TV towers—was insensitive and amateurish. Worse yet, it played into the hands of the US environmental movement, which was just beginning its orchestrated assault on the oil sands industry and could scarcely believe its luck.[37]

In the days that the tailings pond bird deterrent system was first designed, environmental protection was even less of an exact science than it is today, and Syncrude was just starting up the learning curve. The

2008 wildfowl disaster brought the company a cascade of justified public indignation and a reputational dent that would take years to repair. It also rekindled public concern for the reclamation of the oil sands industry's tailings ponds, which covered 220 square kilometres of the region. From outer space, they look tiny compared to the vastness of the boreal forest, and with their white-rimmed "beaches," almost benign:

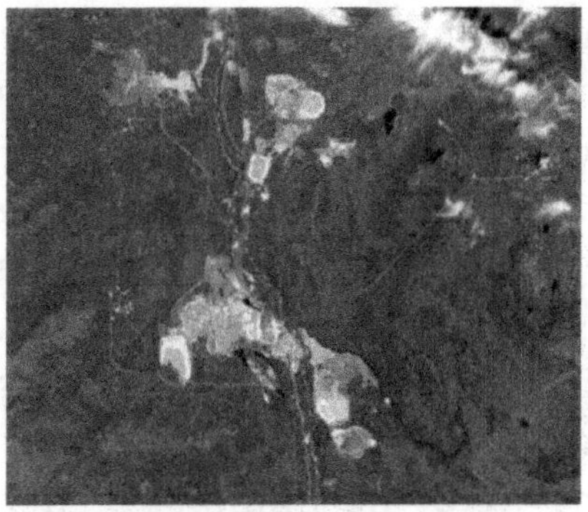

NASA image of tailing ponds. (Photo credit: NASA.)

Closer up, not so much:

Close up aerial view of tailings ponds. (Photo courtesy of David Dodge, Pembina Institute.)

Let's imagine you are developing a scale

to depict industry's degree of toxicity and its harmony with nature. Lake Louise in Banff National Park would be a 10, and the aeration pond of your local human sewage treatment plant would be a 0. If you walked the shore of your average oil sands tailings pond, after a little reclamation had been done, you'd probably give it a 3.

Toxicity and harmony with nature are relative. When the dam containing the tailings pond of the Mount Polley copper mine in British Columbia burst in August, 2014, the cascade of water and tailings wiped out a creek and dumped some 25 billion litres of contaminated materials, including compounds of nickel, arsenic, lead, and copper, into Quesnel Lake. The lake is a source of drinking water and a major spawning ground for sockeye salmon. According to Environment Canada, the tailings pond now contains 326 tonnes of nickel, over 400 tonnes of arsenic, 177 tonnes of lead, and 18,400 tonnes of copper and its compounds, most of which, presumably, is now on the bottom of Quesnel Lake. According to geologists, most of those metals are chemically locked up with rocks and will remain so.

The Mt. Polley tailings dam, after the failure. (Photo credit: Cariboo Regional District.)

What does the failure of the Mount Polley tailings pond say about the relative risk posed by the oil sands' tailings ponds, which are 56 times as large in surface area? Given the amount of residual bitumen they contain, the failure of one of their dams could conceivably dump a large amount of bitumen into the Athabasca basin, not something to be contemplated without concern. It's worth noting that none of the oil sands tailings pond dams have failed or are likely to, given that the oil companies generally sought the best possible civil engineering expertise when designing them.

Having said that, reasonable people hope the tailings ponds will eventually be cleaned up, drained, and converted into self-sustaining wetlands similar to what nature put there in the first place. The question is how long that'll take to happen. Which brings us to the rest of Syncrude's environmental protection report card.

Tailings Pond Reclamation

Goforth, his staff, and Syncrude's first engineers realized early that the mining industry's tailings ponds solved one environmental problem—the need to avoid large withdrawals from the Athabasca River—while creating another: what to do with the dirty water (and residual oil and clays) left over from the hot water extraction process.

The pond reclamation problem was magnified by two things: the sheer volume of water (millions of gallons), clays, and residues that needed to be processed and recycled and the maddening unwillingness of the clays to settle. The original designers of the tailings pond visualized that the clays would settle to the bottom and eventually consolidate so that, some day, the pond could be drained and become a biologically self-sustaining wetland. But early in the pond's life, it became clear that the clays didn't *want* to settle.

Over the past 35 years, Syncrude and other companies with tailings ponds have spent millions testing ways to do get the clays to settle. Flocculants have been tried. Centrifuges have been tried. As yet, no tailings pond has been dried and fully reclaimed, at least to government standards.

One might ask why development even went ahead if there was no proven solution to the problems it might create.. Of course, the same question could be put to most human endeavors including agriculture, the plastics industry, the coal industry, the forest industry, the nuclear power industry, and the conventional oil and gas industry (whose abandoned wells, some of them full of salt water, dot large parts of western North America). Some day the oil sands industry will find a solution. But the wait has been long.

Mine Land Reclamation[38]

The industry's track record in surface-land reclamation is considerably better. The three big oil sands deposits in Alberta (Athabasca, Peace River, and Cold Lake) underlie 142,200 square kilometres of land. Oil sands mining has only disturbed about 4 percent of that land or about 260 square kilometres. About 11 percent of that has been certifiably reclaimed; if land in the process of reclamation were included, the percentage would be much higher.[39]

Wood bison grazing on reclaimed land near Syncrude mine. (Photo credit: Gunter Marx / Alamy Stock Photo.)

The oil sands companies operate major tree and plant nurseries and started reclaiming disturbed area (initially, places like roadsides and cleared areas) almost as soon as mining began.[40]

Imperfect Solution, Failure, or Something in Between?

From the early 1970s until around 2010, Syncrude and the rest of the oil sands industry managed a wide variety of environmental concerns. Early evaluation of the adequacy of industry programs centered primarily on so-called traditional issues like air and water pollution and land disturbance; from the 1990s on, studies shifted to human health, primarily the oil sands' impacts on the Athabasca River and downstream communities north of Fort McMurray, particularly Fort Chipewyan.

Much research was commissioned by companies, universities, and government monitoring agencies. Many scientific studies were mounted. University researchers were widely engaged. Millions were spent in the creation of ongoing research studies. Problems were identified and studied; some were addressed to the satisfaction of any reasonable person, some were not.[41]

It's difficult to separate Syncrude's performance in environmental management from that of the larger oil sands industry. The closest thing we have to an impartial and expert review of industry performance is the study commissioned in 2019 by the Royal Society of Canada, the country's most distinguished multidisciplinary science organization. The RSC commissioned a broad-scale study of the environmental and human-health impacts of the entire oil sands industry in order "to provide a comprehensive, evidence-based assessment of the industry's environmental performance."

Broadly speaking, the RSC's report card gave the industry a B, with several admonitory footnotes. The organization didn't support environmental writers' more extreme claims. The summary of study reads as follows:

"Feasibility of reclamation and adequacy of financial security: Reclamation is not keeping pace with the rate of land disturbance, but research indicates that sustainable uplands reclamation is achievable and should ultimately be able to support traditional land uses. Current practices for obtaining financial security for reclamation liability leave Albertans vulnerable to major financial risks.

"Impacts of oil sands contaminants on downstream residents: There is currently no credible evidence of environmental contaminant exposures from oil sands reaching Fort Chipewyan at levels expected to cause elevated human cancer rates. More monitoring focussed on human contaminant exposures is needed to address First Nations and community concerns.

"Impacts on population health in Wood Buffalo. There is population level evidence that residents of the [regional municipality] experience a range of health indicators consistent with 'boom town' impacts and community infrastructure deficits, which are poorer than those of a comparable Alberta region and provincial averages.

"Impacts on regional water supply. Current industrial water use demands do not threaten the viability of the Athabasca River system if the Water Management Framework developed to protect in-stream, ecosystem needs is fully implemented and enforced.

"Impacts on regional water quality and groundwater quality. Current evidence on water quality impacts on the Athabasca River system suggests that oil sands development activities are not a current threat to aquatic ecosystem viability. However, there are valid concerns about the current Regional Aquatics Monitoring Program (RAMP) that must be addressed. The regional cumulative impact on groundwater quantity and quality has not been assessed.

"**Tailings pond operation and reclamation.** Technologies for improved tailings management are emerging but the rate of improvement has not prevented a growing inventory of tailings ponds. Reclamation and management options for wet landscapes derived from tailings ponds have been researched but are not adequately demonstrated.

"**Impacts on ambient air quality.** The current ambient air quality monitoring data for the region show minimal impacts from oil sands development on regional air quality except for noxious odour emission problems over the past two years. Control of NOx emissions and regional acidification potential remain valid concerns.

"**Impacts on greenhouse gas emissions (GHG).** Progress has been made by the oil sands industry in reducing its direct GHG emission per barrel of bitumen produced. Nonetheless, increasing direct GHG emissions from growing bitumen production creates a major challenge for Canada to meet our international commitments for overall GHG emission reduction that current technology options do not resolve.

"Environmental regulatory performance. The environmental regulatory capacity of the Alberta and Canadian governments does not appear to have kept pace with the rapid growth of the oil sands industry over the past decade. The [Environmental Impact Assessment] process relied upon by decision-makers to determine whether proposed oil sands projects are in the public interest has serious deficiencies in relation to international best practice. Environmental data access for cumulative impact assessment needs to improve."

Predictably, the report did not fully satisfy either environmentalists or the industry, which could be taken as evidence that the reviewers got the subject just about right. The environmental protection measures put in place years before by early industry scientists like Goforth produced benefits and reduced harm. By any reasonable standard, Syncrude—if not the industry as a whole—had tried hard to ameliorate the negative effects of its project while maximizing its social benefits. For the day, that was perhaps as good as it got.

The RSC report was written in the fifth decade of oil sands development, which was just about the time that a game-changer was coming into view, an environmental issue that dwarfed all others, a nemesis that the industry could neither defeat nor ignore.

Climate change.

10. THE BUMPY ROAD TO THE GOLDEN AGE

Between 1980 and 2010—*very* slowly at first, and then with remarkable surge—the modern oil sands industry was born. Its first nemesis—developing workable technology to mine the oil sands and produce large amounts of bitumen—had been wrestled to the ground thanks to Syncrude. The second nemesis—high production costs compared to the rest of industry—was still ubiquitous. Both Syncrude and GCOS struggled with it.

For most of the 1980s, no big new oil sands projects came forward. The world of oil had become violently turbulent. War and revolution convulsed the Middle East. World oil prices, which had soared in the 1970s, crashed and brought down proposed projects. Within a 10-year span, the world oil price went from over $100 a barrel to barely $20 a barrel. At least one oil sands project was cancelled.

After 1990 and until 2010, the oil sands industry grew so rapidly that it became almost synonymous with Canada's petroleum industry. Eighty percent of Canada's oil production was from the oil sands, and if you added in natural gas liquids, the country was producing almost 5 million barrels of petroleum a day and was the fourth-largest producer on Earth.

This transformation was made possible by three developments: the streamlining of the two original mining projects—Syncrude and Suncor —into efficient, profitable enterprises; growing confidence among the major oil companies that oil sands projects could be attractive investments; the pouring of some $200 billion into new production; and the invention of technology that could finally tap the enormous potential of the deep sands.

The 30-year growth of the oil sands into a major global force was a complex process that involved many companies, politicians, inventors, and breakthroughs. This summary of what happened during those years vastly understates how complex and at times perilous the process was. What follows is a highly compressed big picture account built on highlights only.

How It Began

On the strength of Syncrude's completion and successful start-up, it might have seemed to observers that the stars were finally aligning for the building of a great Canadian industry.

Up close, the reality was less promising.

True, Syncrude had been built on time and on budget, more or less.[42] The untested organization that Brent Scott and his team had built successfully started up the Syncrude complex. Synthetic crude oil was flowing in larger and larger amounts, so much so that Syncrude quickly become the largest single oil producer in Alberta and Canada. Someone had finally shown that a world-scale oil sands mining project could be built and could work.

But in the 1980s, the world of oil had become a chaotic place abounding with investment risks. Collapsing oil prices, revolutions, and wars led people—including executives in the oil industry itself—to believe that oil sands development was financial quicksand. Oil price fluctuations and supply interruptions produced a combination of economic stagnation and inflation—"stagflation"—that undermined Western economies.

Syncrude had demonstrated a workable technology, but it hadn't demonstrated profitability, certainly not enough profit to attract the enormous quantities of investment capital needed to continue oil sands development. At least one planned oil sands project that was on the books was cancelled; others were scaled down or postponed.

Simply put, Syncrude's production costs were

too high. In 1981, the world price of oil peaked at about $110 a barrel and then began a terrifying slide, finally hitting bottom at $10–$20 a barrel in 1996 (I've stated it as a range to provide for the fact that oil sands bitumen is sold at a significant discount off the world price). Syncrude absolutely had to increase its efficiency and dramatically drive down its production costs. A cost of $35–$50 a barrel was a formula for shutting down the infant industry and walking away. Scott's successors had to figure out how to significantly reduce costs by reorganizing and reducing the company's workforce and improving its production technology. Once that was accomplished, they could think about how to persuade government to adopt policies that'd improve the economics of new oil sands investments.

Then there could be a golden era for oil sands development. Just maybe.

World Oil Price Fluctuations after 1970. (Figure source: US Department of Energy, https://www.energy.gov/sites/default/files/fotw859.jpg.)

Scott's Successors

After Scott's retirement, the first Syncrude CEO to address the cost question head-on was Ralph Shepherd, a veteran Imperial Oil refinery manager who the owners appointed in 1985 with a mandate to cut costs quickly and deeply.

Shepherd, an electrical engineer by training, was a slight, dapper man who was pleasant enough to anyone who supported his mission. As far as anyone else, he'd stiffly tell them he "wasn't interested in debating." Syncrude had invested heavily in employee communication, by which it

Ralph Shepherd. (Photo courtesy of Syncrude.)

meant two-way dialogue. When I raised this with Shepherd, he said, "I *believe* in employee communication. I am willing to speak to employees at any time."

Shepherd's arrival coincided with the news that Syncrude's production was fetching $10 a barrel, which constituted a return of about 2 percent on the owners' investment. Shepherd made it clear that this state of affairs was unsustainable and that he had a mandate to drive down costs in order to realize a rate of return of 12 percent.

Employees met news of Shepherd's appointment with less than universal enthusiasm. The joke of the period went, "We prayed the owners would send us a Good Shepherd. Instead, they sent us a German shepherd."[43]

Years later, Eric Newell, Shepherd's successor, said, "Ralph once said to me, 'You know the reason we're so good together is that I'm good with things and you're good with people.'"[44]

To his credit, Shepherd's assault on costs wasn't dictatorial or an imposition of top-down draconian cuts. He called for the formation of cost-reduction task forces composed of senior managers in mining, extraction, upgrading, and administration; all teams were asked to identify possible cost savings. Co-operative senior managers who came up with savings were quietly given large bonuses. The task forces suggested measures that included everything from voluntary retirements for people over 55 to just-in-time inventory management and the elimination of free company-supplied coffee at all work stations, which saved $400,000 annually.

Deep cuts were made in the number of company cars provided to managers. Most managers could ride the bus to work, Shepherd ruled. He'd continue to drive his Saab. He didn't believe in leading from the front. He didn't move to Fort McMurray until he'd built the city's largest house, a mansion with a driving range in the basement and a ventilation system designed to keep out the hydrocarbon vapors that triggered his allergies.

In the short term, Shepherd certainly got results: voluntary retirements cut 300 people from the payroll, and at a certain point in time, the company's year-end production cost reached $16.80 a barrel. Most of the cuts were low-hanging fruit; permanent productivity improvements would await major changes in Syncrude's technology, particularly in the mine.

After Shepherd

To be fair, Shepherd wasn't just a single-minded cost cutter. As an Imperial man, he knew an executive was expected to talent spot people with leadership potential, and he did identify and groom the two able men who succeeded him: Newell and Jim Carter. Both were rotated through different management positions to broaden them; Carter was sent to Harvard's intensive executive MBA program. Newell became president in 1989, and Carter became vice president of Operations; he followed Newell into the presidency in 1997.

Eric Newell. (Photo credit: Alberta Trades Hall of Fame, Alberta.ca.)

A measure of their combined effectiveness was that within six years of Newell becoming president, Syncrude shrunk its employee base from 4,700 people to 3,400 and its costs per barrel from $18 to $12.50. It also grew its production from 54 million barrels per year to 74 million barrels, and it did that without any major capital investments—a doubling of productivity.

How it was done was just as important as *what* was done. The two men worked together closely, and over more than a decade, they reshaped the company's operations in a way that honoured Scott's commitment to teamwork and collaboration. In contrast to the downsizing campaigns of many American managements of the day—typically characterized by brutal top-down campaigns to dismiss thousands of employees by executive *diktat*—Newell and Carter worked openly and respectfully with employees to find better ways to utilize their skills, reduce unnecessary costs, and expand production without expanding costs. They made Syncrude into a commercially competitive organization that could attract capital investment and grow. It required forbearance on the part of both management and employees.

The first success factor was a commitment to treat employees as partners in change. When Newell took over, the mood at Syncrude was tense. Newell said, "Our choice was to evolve with the employees right from the outset, talk to them about why we needed to change, involve them in the redesign that would mean fewer jobs. [We knew we had to think about] why . . . employees [would] ever want to do that, put themselves out of a job, so what we said is no, we can't guarantee job security, but we can aim for employment security if you're open [to working with us]."

Carter imported a federal government program called "Bridges" as a pilot to encourage plant administration staff to change careers. A quarter of Syncrude's administrative people were women, and computerization was reducing the need for them. The project encouraged them to consider moving to jobs in mining and maintenance.

"Bridges" wasn't about force-fitting people into jobs they didn't want. Participants were promised that they wouldn't take a pay cut, and in fact, many of them increased their income. Newell later said, "I met with the first 12 women and said, 'Tell me, I want to know the truth: are you doing this even though you hate it and can't imagine going out and working those big trucks or welding on a welding machine, but you have to do it because you're scared shitless about not having a job? Or you're a single mother or whatever the reason is? Tell me.' No one said that, and one woman said, 'Eric, I've done every administrative job in this company, and after three months, I'm brain dead.' One of them said, 'I always wanted to be a welder, and all of a sudden, you put it in my lap. How could I turn that down?'"

Keyano College put on a two-week get-to-know-technology program for the initial intake of women; after that orientation program, the women were offered their choice of jobs to shadow. They were offered an opportunity to try a new job out over a one-month cycle, following which they could elect to keep the job or go back to their old position. According to Newell, no one did:

> It was pretty big change, and we didn't do everything exactly right, but in the end, we were able to manage the workforce down from over 4,700 people to 3,400 people or less, and we never had to lay anyone off. In a world where American corporations often re-engineered their workforces brutally

and without regard to the human cost, this was a major accomplishment.

A large component of Syncrude's change management process was face-to-face communication between senior management and the ranks, something Shepherd never really grasped. Here's how Newell later described it:

> I had the company divided into 70 different units, and I had to go talk to every one of those 70 units at least once a year, and then the other thing we did is every 18 months, we had all-employee meetings, and it would take us 19 meetings at two to three hours a crack to cover all the employees, and they would all get time off work to bus in, and we [would] do them around the clock. Jim Carter and I would usually lead off and give a 20-minute talk about the state of the nation, and we did a lot of up-front work trying to find out what we thought were issues on employees minds and address [them], but then they all sat there. It was comical at times, but . . . they all had cards at their seats, so they'd write in these questions anonymously, and Jim and I would have to stand up there for two hours, reading these questions out and answering them, and boy, I'll tell you, we

learned what was on employees' minds; every time, we learned something new. It was one of the most valuable things we ever did, and then of course, you have to follow through and deliver on it. So I think employees felt engaged. The other thing we did is we put in a gain sharing program, which I always congratulate my owners for, because . . . [the owners' organizations in Calgary] were laying people off left and right, and so, for the owners to approve a gain-sharing program was tough. But I knew it was the right thing, and it was the employees that decided this, that we always use to form employee task forces to work key issues, and it was them that said the secretary should be paid the same amount as the CEO on the gain share.[45]

The second success factor was intensive leadership training to help make team management work at all levels. Syncrude put more than 1,100 of the company's supervisors and future supervisors through it; at the end of every session, Newell and Carter met with the groups for frank discussions. As Newell put it, "Boy, I'll tell you, they didn't mince their words. . . . We used that vehicle to also develop a shared vision of where we're going and what are the values we're going to run our company by and what principles are we never going to compromise. . . . We also put more than 80 percent of our people through diversity training." At the time in Canada, that much diversity training was quite unusual.

Newell was a warm personality and good listener who could be candid without being overbearing. He had excellent people skills and a grasp of the importance of politics in Syncrude's relationship with the world. In the 1990s, he emerged as a likeable spokesperson not just for Syncrude but for the industry.

In addition to his internal accomplishments, Newell saw the need to organize a successful lobby for better federal and Alberta government royalty and revenue sharing policies. Newell was the first CEO since Spragins to fully realize that the oil sands would never be developed to its full potential without supportive public policy (i.e.., provincial and federal tax and royalty agreements). He also understood the politics of achieving that end. The oil industry wasn't universally liked in Canada, let alone loved, and no single oil company—not even the respected Imperial Oil—had the political capital to influence public opinion and move politicians. What was needed was a broader coalition.

In 1995, as president of the Alberta Chamber of Resources, Newell spearheaded the creation of a National Oil Sands Task Force that encouraged government to develop a comprehensive energy strategy for Canada. That strategy unlocked more than $50 billion of private investment in oil sands projects that completed the transition from start-up to maturity by 2015.

Newell devoted major time and energy to becoming the de facto voice of the oil sands, and he spent years on the road dialoguing with politicians, civil servants, and editorialists across the country. At one point, he gave more than 50 major speeches a year. He worked with the Alberta Chamber of Resources—at the time, a less than powerful interest group—to create the National Oil Sands Task Force in 1993. Over a two-year period, the task force involved 35 organizations across Canada in developing an energy vision for Canada and the oil sands. Said Newell:

> Instead of just talking about it and how we could improve the profits of Imperial or Syncrude or Suncor or whatever, which a lot of people don't really care about . . . [we talked about things they did care about,] like jobs, royalty and tax payments, and [maintaining] the social systems that we value so highly as Canadians.

Supported by the solid work of two Syncrude staff—Phil LaChambre and A.W. Hyndman—Newell sold key players in the federal and Alberta governments (headed at the time Brian

Mulroney and Ralph Klein respectively) on the proposition that appropriate tax and royalty regimes would unlock a flood of new investment in the oil sands—indeed, the figure he used was between $21 billion and $25 billion—to say nothing of job creation across Canada.

"I used to get a cold sweat," said Newell, "because at that time, we were investing zero dollars, so people would look at you like you were maybe smoking something funny. Once we got everybody excited about what the potential was, it was amazing how relatively easy it was to figure out how to knock down those barriers to development."

In June 1996, Prime Minister Jean Chrétien and Minister of Natural Resources Anne McLellan, and Alberta Energy Minister Pat Black came to Fort McMurray and announced that there were $6 billion dollars of projects that were ready to go; all of them had signed a so-called Declaration of Opportunity. Following that declaration—which visualized possible growth over 25 years—it took only eight years (and a $100 billion of investment) to expand oil sands production five times. Development became a locomotive.

Reinventing the Technology

During his 28-year career with Syncrude, Carter revolutionized the company by dramatically redesigning its approach to mining.

A native of Scotland, he grew up in Prince Edward Island, trained as a mining engineer in Nova Scotia, and worked in the iron ore mining industry and the Alberta coal industry before coming to Syncrude in 1979. He came to be seen as a mining engineer's mining engineer.

Jim Carter. (Photo courtesy of Syncrude.)

Early in his career, Carter recognized the need for economies of scale in the oil sands, so he made the controversial decision to switch from draglines and bucket-wheels to a trucks-and-shovels form of mining; this process took years, and Carter worked hard with mining equipment manufacturers to prompt development of the world's largest electric-powered shovels and the world's largest 400-tonne capacity trucks. He was part of the hydrotransport solution (more on that shortly). Together, these changes in materials handling reduced the energy required to produce a barrel of bitumen by 40 percent, with economic and environmental benefits to boot.

During Carter's 1997–2007 term as Syncrude president, the company's production rose from approximately 200,000 barrels per day to 350,000, and its market capitalization grew from $3 billion to approximately $45 billion. This summary doesn't do justice to Carter's full contribution, let alone that of hundreds of managers and employees working in teams over many years. The development of all the technology changes Syncrude needed to reach its full potential as an oil producer took thousands of people hundreds of person-years at all levels of the organization and some 18 years to wrestle all the problems to the ground, but wrestle them Newell and Carter did.

The greatest focus of the changes was on the mining and extraction parts of the operation. Carter characterized the main goals as achieving safe reliable operations, reducing chronic failures, putting more throughput through the mining and extraction parts of the operation, increasing fluid coker run lengths and liquid volume yield, and motivating and realigning the workforce. His was a complex vision.

The shortcomings of the original mine design —draglines and bucket-wheel reclaimers to mine and stockpile the oil sand and conveyor belts to move it to the extraction plant—had become apparent. Draglines and bucket-wheels had been invented primarily to mine soft coal. Both types of machines—the bucket-wheels built in Germany and the draglines in the US Midwest —were expensive to manufacture and even more expensive to maintain. Both had major downtime for repairs and were costly to fix. Draglines had difficulty moving to new locations in the summer when the soil was soft. Syncrude had 300 kilometres of conveyors to move oil sand from the mine to the extraction plant. They were expensive to move when the sand beneath them had to be accessed, and they were prone to breaking down. Plus, they'd be far too expensive to operate if the company wanted to expand to new mining areas.

The conveyor problem was solved with the invention of hydrotransport, which was the process of crushing oil sand lumps in Syncrude-designed crushers, mixing the crushed lumps with lukewarm water, and pumping the slurry through a pipeline to the distant extraction plant. Hydrotransport solved several problems: it eliminated the breakdowns to which conveyors were subject; it reduced the cost of transporting the oil sand long distances; it reduced the temperature of the water needed for extraction; and it improved the efficient separation of the bitumen from the oil sand.

Hydrotransport made it possible for Syncrude to open a whole new oil sand mine, Aurora, in 2000, and made it possible for Syncrude to run at higher output for another 60–70 years. Hydrotransport was a team effort that incorporated the insights of three people at Syncrude's R&D laboratory in Edmonton —Tony Leung, Waldemar Maciejewski, and George Cymerman—as well as people from Syncrude's mining operation and researchers at the Saskatchewan Research Council, the Alberta Research Council, and the Federal Upgrading

Laboratory. Victory has a thousand parents; defeat is an orphan.

Syncrude gifted the idea of hydrotransport to GCOS.

Suncor's Makeover: The Saga of Rick George

Rick George. (Photo courtesy of the Canadian Petroleum Hall of Fame.)

While Newell and Carter were making Syncrude bigger and more profitable, another executive, Rick George, was transforming Suncor, which was at that point the sick man of the sands, into the biggest force in the oil sands industry and a major force in the country's overall industry as well.

George's story, told in his book *Sun Rise: Suncor, the Oil Sands and the Future of Energy*, demonstrated his considerable talents both as an oil industry transformer and entrepreneur. While the story's been told elsewhere, its highlights are worth recounting since they played such a major role in changing the industry.

George was CEO of Suncor from 1991 to 2012, a period in which he overhauled GCOS's struggling, money-losing operations, modernized production and processing techniques, and invested billions to expand the company into a business that now pumps nearly half a million barrels a day. He converted Suncor from a subsidiary of the US company built by J. Howard Pew into a Canadian company with its own Canadian shareholder base and then a corporate giant incorporating Petro-Canada.

His technology makeover of GCOS—in the mine at least—was similar to Syncrude's. He abandoned GCOS's original mining technology of German-built bucket-wheel excavators and its replacement with a trucks-and-shovels scheme. This took place at a time when GCOS was spending $20 to produce a barrel of oil that sold for $12, a situation that obviously couldn't continue for long without a disastrous ending.

The other part was his engineering of Suncor's spectacular $19 billion takeover of Petro-Canada, which more than doubled the company's size and made it into a fully integrated giant (refining, marketing, and production). Described as "both a contrarian and an optimist," George often made multi-billion-dollar moves despite deep skepticism within the industry. His $2.8-billion expansion into the oil sands helped fuel massive growth in Canada's most valuable natural resource. And his $19-billion merger with Petro-Canada in 2009 turned Suncor into a true global player in the international industry.

Since the days of J. Howard, Sun Oil had been an American-owned, Philadelphia-based company with its own shareholder base and history. In the 1980s, Robert McClements succeeded J. Howard as president and CEO of

Sun Oil. McClements reached into Sun's North Sea oil operation and asked George, who was in charge of it (and had built Europe's first purpose-built offshore production platform), to take over Suncor's operations in Canada (of which GCOS was the largest part).

George became president in 1990, when Suncor's principal assets were GCOS, a small refinery, and a service station network in Ontario—an unlikely platform for major growth and prosperity! He was an opportunist (in the good sense of the word: a person who recognizes an opportunity where others don't and has the skill and courage to seize it). In 1991–1992, there was a recession. Suncor's parent company in Philadelphia had debt problems, and so did the Government of Ontario, which at that point owned 75 percent of Suncor Canada. George persuaded both to sell the company to Canadian public shareholders. This transformed Suncor Canada into an independent company headquartered in Toronto.

Then George focused on overhauling Suncor's struggling mining operation, moving it from bucket-wheel excavators to a trucks-and-shovels operation like Syncrude's. He also improved GCOS's processing plant and in 1998 applied for a major production expansion to 210,000 barrels a day. That made GCOS the second largest oil sands producer.

Then, in 2009—after an earlier failed attempt

blocked by Petro-Canada's largely government-appointed board—George stickhandled a $19.1 billion bid to take over Petro-Canada. In this, he displayed an entrepreneurial finesse that very few oil industry executives demonstrated, at least among the large companies (so-called junior oil and gas company CEOs were a different breed). The market responded, and Suncor was catapulted into a major oil company.

George was a practical student of economics, human behaviour, and energy. He had a much shrewder understanding of communication and advocacy than most oil company CEOs and a way of communicating Suncor's point of view that radiated sincerity and openness. And he was also a natural when it came to what Syncrude called "team management." This enabled him to enlist employees, shareholders, business partners, and governments—not to mention citizens at large—in his mission. Sadly, he died in 2018 from acute myeloid leukemia. He was 67.

During his administration, Suncor became the largest shareholder of Syncrude, a step that, as this is being written, entails Syncrude's absorption into the Suncor "family." An interesting question is why Syncrude didn't continue on its path to becoming the dominant force in the oil sands industry. It had a head start, plenty of skilled people, and a tonne of intellectual capital. Until 2009, it was well managed by a sequence of leaders. So what happened? The answer is that it was hobbled by its joint venture structure and its ownership by other companies. In truth, its owners never intended it to become an independent force. As the operating arm of a corporate joint venture, its shareholders were other oil companies; major strategic and investment decisions were made by, essentially, a committee of owners, each reporting to distant head offices anywhere from Calgary to Toronto to the United States, Europe, and even China.

After Carter retired, the owners appointed a series of no-name executives who came and went without much of a trace. And a hugely expanded Suncor, which bought out other Syncrude shareholders, became the majority Syncrude shareholders and its eventual voice.

At Last, the Deep Oil Sands Take Centre Stage

For the first 30 years of modern development, starting with the commencement of GCOS in 1962, the oil sands industry was driven by giant mining projects. In the 1990s, that was about to change.

Geologists had long known that the largest part of the oil sands—the generally accepted estimate was 80 percent—was too deeply buried to be mined economically. The cost of moving the overburden would vastly exceed the value of the oil uncovered. In these deep sands—500 to 2,500 feet below the soil and subsoil—a means had to be found to separate the bitumen from the oil sands in situ and pump it to the surface.

In the earliest days of the sands, there were many failed attempts to do this. Simply drilling wells into a given formation didn't work; the sticky bitumen wouldn't flow and couldn't be pumped. Various schemes were tried to heat the bitumen, but none of them worked. In the 1970s, Amoco Petroleum developed a technology that used down-hole combustion and waterflooding (called "combination of forward combustion and waterflooding"), and it looked promising until workers found that the fire in the hole melted the drill casings. Others hypothesized that the deep oil sands could be heated by microwaves.

In the 1950s, an American petroleum engineer proposed Project Cauldron—the detonation of small atomic bombs underground. The idea was to use nuclear explosions to create a huge superheated chamber in which the cracked bitumen would collect and could be pumped to the surface by drilling. Promoters of the idea were unable to answer obvious questions, like what would happen if the blast ruptured the surface and sent nuclear fallout for hundreds of miles. The idea also ran up against the movement to ban nuclear testing, and it never went beyond the drawing board.

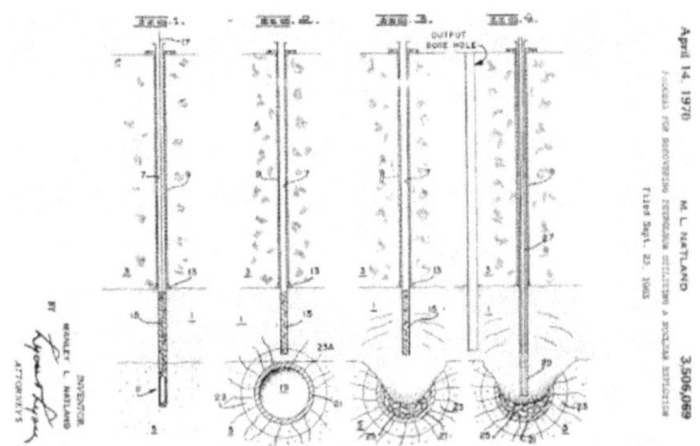

"Project Cauldron" conceptual design (Image source: Patent.)

Peter Lougheed's Contributions to the Oil Sands Breakthrough

P.E. Trudeau (left) with Peter Lougheed (right). (Photo courtesy of the Provincial Archives of Alberta, J3672.2.)

During his time as Alberta premier (1971–1985), Peter Lougheed made three major contributions to the development of the oil sands.

The first was his facilitation of a new oil sands royalty scheme that balanced the public interest with the risk management concerns of investors. This was a royalty based on the industry's deemed net profits. Prior to Syncrude, Alberta based its oil sand royalties on production volume (e.g., 12 percent of every barrel produced), the same approach the province used for the conventional industry. In 1973, Syncrude's owners were able to make the case that given the extraordinarily high capital cost of an oil sands plant and its high risk, a volume-based royalty would make it impossible to finance megaprojects. Instead, they proposed a net profit royalty system in which royalties would be paid once initial capital costs had been recovered. The royalty would be 50 percent of net profits after payback of initial capital cost. With some updating, this concept has continued in force since.

Lougheed's second accomplishment was the creation of the Alberta Heritage Trust Fund, which was intended to set aside a portion of oil sands revenues for future generations. Sadly, Lougheed's successors had little idea of how to invest the funds wisely in potential next-generation projects; most of the projects they did invest in failed, including a magnesium production plant and a regional bank. A 2015 paper called "The Siren Song of Economic Diversification: Alberta's Legacy Loss" by F.L. Ted Morton and Meredith McDonald explores the subject more fully. Following that, successive Albertan Conservative governments succumbed to the temptation to spend much of the oil sands royalties on government expenses, and the Heritage Fund never became the funder of Alberta's future that it could've been. Good concept. Lousy execution.

Lougheed's third accomplishment was the creation in 1974 of a government-funded R&D enterprise called the Alberta Oil Sands Technology Research Authority (AOSTRA).

Funded with an initial capitalization of $100 million, AOSTRA's mission was to foster the development of breakthrough oil sands technologies. And in the late 1990s, with AOSTRA's support, the breakthrough happened. The key figures were two ex-Imperial Oil chemical engineers, Clem Bowman and Dr. Roger Butler. Bowman was AOSTRA's first director, and Butler was the driving force behind its major technological breakthrough: steam-assisted gravity drainage, or SAGD.

Butler graduated with a PhD in chemical engineering from Imperial College in England and in 1955 joined Imperial Oil, where he worked on heavy oil and oil sands production technology for 27 years. In collaboration with Bowman, he was instrumental in

Roger Butler. (Photo courtesy of the Canadian Petroleum Hall of Fame.)

developing Imperial's cyclic steam stimulation method—or "huff'n puff," as it was described in the industry—of producing heavy oil from the Cold Lake formation.

It was SAGD that brought the deep oil sands within range of large, sustainable bitumen production. Using computer-controlled horizontal drilling, it allowed for the creation of a very large-bore vertical tunnel into oil sands formations. From this tunnel, two sets of horizontal tunnels would be drilled: one into formation and a second about 5 metres below. Super-high temperature steam would be pumped through the upper tunnel; it'd melt the bitumen, which would then run down into the lower tunnel and be collected and pumped to the surface. AOSTRA decided to fund the pilot project to demonstrate the technology after sending a delegation to Russia in November 1976 on a hair-raising visit to a primitive oil mine site in Siberia.

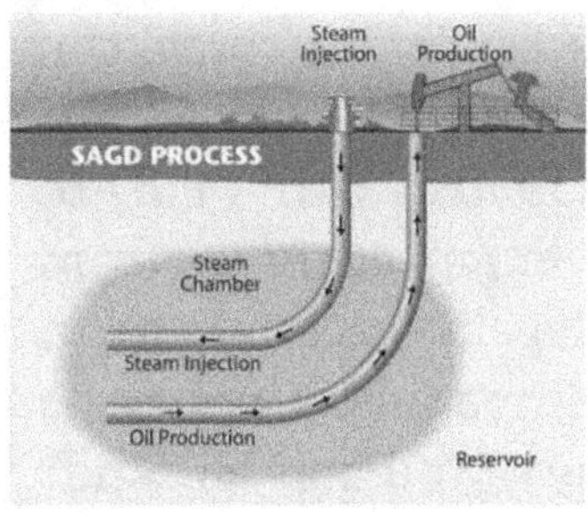

SAGD Process. (Figure source: The Canadian Centre for Energy Information.)

SAGD revolutionized the oil industry because it made large-scale production from the deeply buried oil sands at a dramatically cheaper per barrel, and it did that without the huge capital and operating costs of mining. Thus, in theory at least, it enabled a rapid expansion of oil sands production at a significantly lower price.

It had another huge benefit: it could be scaled down into small, affordable operations, which could then be linked to major production complexes. This opened an opportunity for nimble, smaller companies to get into the oil sands business. In situ producers eventually included companies like Cenovus, Japan Canada

Oil Sands, Nexen, Husky Energy, and Devon Energy as well as majors like Conoco Phillips, Suncor, and Shell.

On the face of it, in situ had another advantage over mining: a dramatically reduced environmental impact, at least on the Earth's surface. No more massive open-pit mines and their immense tailings ponds. With in situ, the separation of bitumen from the sands would take place deep underground instead of in a massive extraction plant on the surface. The water would be cleaned, reheated, and recycled far below the surface, and the waste would stay buried.

Of course, things are never that simple. In situ raised at least three environmental issues:

- it needed to draw large amounts of fresh surface water from lakes and rivers to run the process,
- it carried the potential risk of groundwater contamination, and
- it needed to burn large amounts of natural gas to make steam, a prefiguring of increased greenhouse gas emissions.

A Landslide of Investment

By the middle of the 1990s, these three factors—the maturation of Syncrude and Suncor into profitability, the provision of strong incentives for new investment in the oil sands via innovative government revenue-sharing systems, and the technological breakthrough of SAG-D and cyclic steam stimulation at Cold Lake—fuelled an explosive growth in oil sands production of both synthetic crude and bitumen. In situ projects charged forward in the three major oil sands regions. At the G8 conference in St. Petersburg in 2006, Prime Minister Stephen Harper announced his intention to see Canada become an energy superpower. By 2012, Alberta produced about 1.9 billion barrels of crude bitumen a day from the three deposits, over and above the mining sector's production of synthetic crude oil.

By 2020, the Alberta government estimated that the oil sands had generated $325 billion of capital investment and accounted for three quarters of Canada's oil production of almost 4 million barrels per day (5 million barrels if natural gas liquids were counted). Oil exports had become an important revenue earner in the Canadian economy. Canada was the fourth or fifth largest oil producer on Earth. By 2020, producing more than 50 percent of Alberta's total oil production, in situ was the international success story of the Canadian petroleum industry and was remaking the face of the oil sands.

Canada—or more properly, Alberta—was becoming a force in the oil world. In 2019, Canada was the largest foreign supplier of crude oil to the United States, accounting for

48 percent of total US crude oil imports and for 22 percent of US refinery crude oil intake. Canada exported 3.7 million barrels per day to the United States in 2019, 98% of all Canadian crude oil exports. When natural gas liquids were added in, Canada was producing nearly 5 million barrels of oil a day, making it a significant world player in oil production.

Social and Economic Impacts across Canada

The explosive development of the oil sands between 1985 and 2015 engineered an astonishing movement of people seeking opportunity and a major transfer of wealth within Alberta and between Alberta and other Canadian provinces. People outside our oil producing regions sometimes lost sight of the benefits that this brought the whole country.

Not only did the capital spending of the oil industry generate enormous income for suppliers across the country, skilled workers flooded into Alberta from every province, often from depressed communities. Entire towns in the Atlantic provinces, their local economies in decline, witnessed a virtual rebirth from the dollars shipped home from the oil sands by enterprising men and women. A rising tide lifts all boats, and many of the boats were in coastal towns in the Atlantic provinces, not to mention in Indigenous communities across northern Alberta. The benefits of oil sands spending spread to every province.

Oilsands Supply Chain. (Figure source: CAPP.ca, How the oil sands benefit all of Canada - Context Magazine by CAPP.)

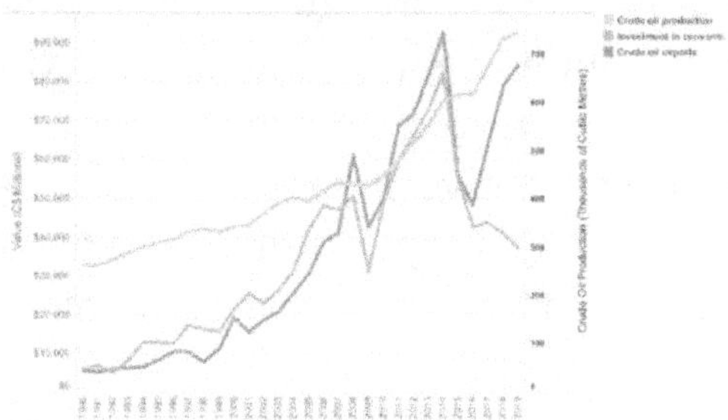

Canadian Trade of Crude Oil. (Figure source: Canada Energy Regulator: Canadian Crude Oil Exports: A 30 Year Review.)

All that increased oil production had to go somewhere, and it did: to the United States. It could also have gone to Asia. There were available markets. But to get either to refineries in the US Gulf Coast or to Asia, it needed to be shipped, either by tanker or pipeline. The extraordinary, rapid growth of the in situ industry created several bitumen problems and brought the industry, Alberta, and Canada into the path of the oncoming locomotive of climate change.

The fact that a growing proportion of oil sands production was in the form of bitumen created a major problem that undermined the possibility of future growth. Unlike oil sands–mining companies that upgrade their bitumen into synthetic crude, in situ producers generally produce raw bitumen and then dilute it with petroleum solvents to make it suitable for pipelines. Diluted bitumen or "dilbit" was quickly demonized by environmental groups who charged that it had a sky-high greenhouse gas component and was a toxic risk to communities anywhere near pipelines. It was also alleged that spilled dilbit is much nastier stuff to clean up after a pipeline break or tanker spill.

Producers of bitumen from in situ plants targeted large, modernized, and new refineries in the US Midwest and Gulf Coast as well as in Asia that could take bitumen and process it for market. To get the diluted bitumen to these refineries required vast new pipelines, not to mention marine terminals, both of which had to be approved by governments vulnerable to environmentalist pressure.

As we will explore in the next chapter, this push deeper into the American oil market took the industry into politically hazardous waters. Market access would require the building of new US pipelines, and the environmental movement painted those pipelines as both contributors to global warming and threats to American health and safety. Organizing to stop what they called "dirty oil sands pipelines" would become an easy environmentalist tactic.

By 2015, it was becoming clear that the days of rapid oil sand expansion—or *any* expansion—were nearing their end.

PART 3: HOW IT ENDS

11. THE CREATION OF THE MIND FIELD

At the end of the 20th century, Canada's oil sands industry appeared to be entering a golden age. Powered by technological change and stimulative government policies, in situ production from the deep oil sands was skyrocketing, and industry as well as the Alberta and Canadian governments were supporting massive exports of synthetic crude and bitumen to both the United States and Asia. The oil sands had become the base of a massive Canadian oil export industry.

Neither the industry nor the governments of Alberta and Canada recognized that the drive for new export markets would expose them to grave new political risks that arose from the intersection of two powerful, even revolutionary trends in Western society:

- the growth of the Mind Field, a change

in public consciousness brought on by the growth of digital social media and its exploitation by political forces; and,

- the birth of the climate change movement.

To understand how the possible future of the Canadian oil industry began to come into question after 2010, we first need to understand the profound change in Western consciousness that was brought on by the explosive growth of digital communication and its impact on the Mind Field. Then we need to understand how the climate change debate exploded into world politics and how it played out in the Mind Field.

What Is the Mind Field?

The Mind Field isn't a place. You won't find it on a map. It exists in our minds and our shared thoughts. It's a metaphor for the space where our collective perceptions are shaped and reside. The consciousness of our society—our shared images and imaginings, everything embraced by what we used to call public opinion—is shaped in it. It's where reputations live. It's also where the forces threatening the future of the Canadian oil industry came into play.

You're in the Mind Field now, as you read these words.

The Mind Field grew out of the interaction of language and communication technology. As the two grew, the Mind Field grew with them.

At the beginning of human civilization, there were many Mind Fields, each one like a small bubble enclosing a family, clan, or village. In those days, each Mind Field was small and local because for most of human history, most people spent their lives in families, small clans, and isolated settlements. People's perceptions and opinions were shaped by their family, their neighbours, their priest or headman, and their direct encounters.

As oral and written communication developed and settlements became towns, cities, kingdoms, and nation-states, those bubbles grew and merged. Mass education, mass travel, and the invention of communication technologies like moveable type and wired and wireless communication propelled the expansion of those bubbles. The bubbles came to enclose regions (think William Aberhart on the radio, preaching the gospel to people on the Prairies), then whole countries (think FDR's radio-broadcast fireside chats with all Americans on the radio).

In the 1960s, Marshall McLuhan coined the concept of the global village. At the time of his writing, it was more an idea than a practical reality. What made it a reality was the creation of the internet and all the digital communication technologies, including social networking, that it fostered. The world wide web, Google, Facebook, and other developments shrunk space, connected people, and brought the global village or something like it into being.

The rapidly increasing reach of communication technologies expanded the Mind Field into global proportions, a place where millions and then billions of people acquire much of their information and collaborate on everything from hairstyles to product preferences to religion and politics. While the Mind Field is vast, it's segmented by language, cultural barriers, and, in places, totalitarian governments. Early thoughts that the internet might create a kind of global community have been replaced with the realization that there are multiple Mind Fields operating simultaneously in different parts of the world or within individual societies. China has almost a billion internet users, and the government employs two million people to monitor Internet and social media communication and use it to control behaviour and expression.

The penetration of digital communication technologies into every aspect of our lives has been both blessing and curse. It's fostered universally accessible search engines like Google and knowledge bases like Wikipedia. It's promoted distance education, opening the minds of millions of children in villages and liberating millions of people from superstition and ignorance. It's put powerful research tools into the hands of millions of people and given individuals a voice. It's helped bring down oppressive governments. It's helped free political prisoners around the world and wrongfully convicted people on death row. It's revolutionized the music, film, and publishing industries. It's made it possible for thousands of previously unknown musicians to disseminate their music around the world and thousands of writers to publish their work (including the e-book you're reading now.)

The social media part of the Mind Field is a primary driver of increased tribalism, irrationality, and campaigns of mass deception. Social media has made it possible for millions of people to pursue the truth; many times, it's also made it easy for a lie to travel around the world three times before the truth can even get its trousers on. Social media has made it possible both for Edward Snowden to educate millions of people about secret government wrongdoings and for Republicans to convince millions of Americans that Democrats "stole" the 2020 election from Donald Trump and that the rioters who marched on the US Capitol were actually Democrats pretending to be Trump supporters.

At its worst, digital communication has led too many people to spend far too much of their time and headspace in a tabloid universe in which Elvis is alive, John Kennedy is living on a remote tropical resort, NASA faked the Apollo moon landings, and 9/11 was triggered by

Jewish scientists operating a laser on the moon.

Many catastrophic changes in world history have been ambivalent; the digital revolution is no exception. The Black Plagues of the Middle Ages killed between 30 and 50 percent of the population of Europe but substantially improved living standards and economic opportunities for the survivors and their descendants, who generally lived to much older ages.[46]

In the 21st century, all of this has led to a pollution of public discourse with unfortunate consequences, beginning with the erosion of journalistic standards and research capability in the mass media, which, in the words of one analyst, "has fallen into the immediacy and sensationalism of the digital world."

The hollowing-out and deterioration of traditional mainstream (print) news media is one benchmark of this degeneration that's taken place across America, Britain, and Canada. Witness the degree to which three of the leading mainstream daily newspapers—*The New York Times, The Guardian*, and *The Globe and Mail*—have openly exchanged their traditional detachment for the role of advocate for causes including diversity, climate change, and environmental justice.[47]

Whether the deterioration of mainstream journalistic standards has been caused by the rise of digital media or is the effect of it, the economic decline and hollowing out of the mainstream media is a fact of life. Across the Western world, print newspapers are ghosts of their formal selves, newspaper closures are widespread, and journalistic employment is probably less than half of what it was in the 1960s. Social media has both shouldered aside the mainstream media as a cultural influence and at the same infected it with its spirit of partisanship and the unconscionable pursuit of clicks and eyeballs with flaming headlines, half-truth charges, and unproven allegations.

These shifts have promoted a growing distrust of government and racial, ethnic, and national others as well as a growing, fervent belief in fake news, delusions, and preposterous conspiracy theories, to say nothing of the rapid growth of tribal thinking in US and European politics. I can't forget a conversation I had in 2016 with a very nice woman in an Arizona supermarket. Trump, she said, would soon "clean up the mess in Washington" and send Hillary Clinton to jail, since it was well known that Clinton and her supporters had murdered 42 people who'd gotten in her way.

"How do you know that?" I asked as politely as possible.

"I guess you don't get Fox News up in Canada," she said.

As you read this, countless people and organizations are using the Mind Field to try to influence what you think of their ideas, feelings, leaders, public policy proposals, and products (yes, I'm one of them). Add to that the hundreds of thousands, if not millions, of people who make their living in advertising, public relations, journalism, management consulting, politics and lobbying, not to mention bloggers, Twitterers, fervent believers of every kind, and ordinary human beings just speaking their mind. In short, a significant part of the human race.

What all these people have in common is a desire to shape your thinking about a political party, a politician, a product, an organization, an individual . . . or just an idea.

The insubstantiality of ideas and concepts in the Mind Field doesn't mean that they lack force in our lives. In *Margin Call*, a film about the 2008 financial crash, a financier tells a man who is depressed about all his losses: "It's just money . . . pieces of paper with people's pictures on it so that we don't have to kill each other just to get something to eat."

* * *

To better understand how the Mind Field influences thinking, please join me in a thought experiment. Observe the picture that comes into your mind—and your feelings and reactions—when I mention the name of someone you've probably never met or interacted with in any way.

Here it is: "Bill Gates."

What thoughts, feelings, or mental pictures came to your mind?

Possibly none. Possibly the picture of a super-bright computer nerd who just managed to build one of the world's great corporations, sometimes by destroying competitors. Or the picture of a visionary philanthropist who spends billions of dollars expanding public health systems around the world, fighting Ebola and malaria and tropical diseases. Or the picture of a sinister underwriter of genetically modified crops in Africa and Asia.

Now consider this: neither you or I have ever met Gates nor questioned him or seriously examined his life and beliefs. Neither of us has any basis for thinking that we understand him or can place him somewhere on the scale of ethical behaviour.

Each of our conceptions of Bill Gates is simply an idea that came to us through the Mind Field. None of them is the real Bill Gates, a man no doubt as complex as anyone else.

This is why we need to be careful and provisional about our assessment of people, issues, and the world in general, as most of them are strongly influenced by the Mind Field. Other than people who are close to us (and sometimes not even them!), most of what we think we know about other people, public figures, political parties, and whatnot is simply an opinion, a residue or computation of what we've heard or read from others. In other words, a thought shaped by the Mind Field.

If we want to understand the role that the Mind Field has played in the life of the petroleum industry, we need to consider how it works. It is, after all, the most powerful influence on all the reputations that reside in our brains.

What the Mind Field Is and Isn't

All forms of reputational influencing —journalism, marketing, advertising, public relations, political and ideological activism, politics, and lobbying—shape and operate within the Mind Field.

You enter it every time you sign into Facebook, do a Google search, read a newspaper or book, talk with friends, watch television, go to the movies, or turn on your computer. You're in it now, as you read these words. It's beginning to shape the way you perceive this book, as well as your perception of me, my ideas, the work I do, and the world in which I do it.

A widely held but misleading idea is that the Mind Field is a kind of court of public opinion. Courts are supposed to be places in which the innocent have nothing to fear because our legal system has procedures to guarantee fair play. Therefore, you might assume that a court of public opinion is a place where reputations can expect fair treatment.

This analogy is dangerously false.

In Western countries at least, courts are guided by laws and precedents, and they have detailed, time-tested processes for establishing the facts and applying the law. They have judges and juries guided by precedents, rules of evidence, and procedures. They have mechanisms and practices designed to minimize the role of emotion in the justice system.

The Mind Field not only lacks any mechanism to minimize emotion, but large parts of it are mostly *about* emotion. It's like a raucous town meeting with thousands or millions of participants, no rules of procedure, and no moderator. A place where the loudest and most persistent voices get the most time and attention.

In a Platonically perfect world, the Mind Field would have a code of behaviour to promote civilized discussion. There'd be rules of engagement to minimize emotionalism and unreason. There'd be enforceable rules against the use of misleading statistics, invented facts, appeals to prejudice, and reputational murder by innuendo.

Unfortunately, that's not the world in which we live.

Apart from the laws regarding libel and defamation—both of which operate only in a very limited way in the Mind Field—the biggest factor determining what's allowed, what succeeds, and what doesn't in the Mind Field is its *culture*. Its culture includes social controls like custom, taste, taboos, and appropriateness. Its culture and values *very* loosely set bounds on what can and can't be successfully advocated.

The fact that the Mind Field is almost lawless doesn't mean that *every* kind of advocacy and persuasion work on it. You need to become fully conscious of who's listening, what they're hearing, and what their cultural suppositions are because that'll colour and shape not what you're saying but what people think they're hearing. You also need to be aware of the filters through which people hear what they want to hear.

It's important to understand these social controls, since they operate to some degree across the Mind Field (the Western part of it at least) and can have a bearing on the success or failure of attempts to shape opinions.

Mind Field Culture

Trying to define Mind Field culture is like trying to measure the dimensions of a fogbank. Nevertheless, here's my sense of it.

Over the last half century, at least in what we used to call the Western world, the values of the Mind Field have shifted in the direction of secularism, greenness, and liberalism. The drivers of this greening and liberalizing trend have included:

- the explosive spread of self-expression

(of which social networking is the major manifestation);

- a steep decline in public trust in traditional institutions, government especially;
- the various liberation movements (racial, gender, etc.);
- the spread of the civil society movement, with its canon of universal beliefs like transparency, accountability, inclusiveness, and stakeholder consultation; and most importantly,
- the rise and omnipresence of environmental consciousness.

I would argue that since the 1960s, these forces have strongly shifted the culture of the Mind Field toward values that are secular, humanistic, culturally liberal, and green. Arguably, this process has gone furthest in Western Europe, covered much less distance in the United States, and is somewhere in between in Canada.

How does that culture translate into the Mind Field's rules of engagement? There's no guidebook on how to participate in the Mind Field. Not that some aren't trying to create one. Nisbet documented a 2019 *Guardian* management memo to its journalists stating that henceforth, "climate change" would be referred to exclusively as "climate emergency, crisis or breakdown," and "climate science denier" or "climate denier" is to be used instead of "climate skeptic." And many universities have allowed "politically correct" rules on speech, hiring, and conduct to make numerous, deep inroads into their governance. At this moment, there's no universal statement of them that has the force of law, but they increasingly permeate Western culture and sometimes take the form of a subtle and not-so-subtle group think.

When the Fossil Fuel Industry Realized It Was (Also) Operating in the Mind Field

In the first decades of the Mind Field's expression, many leaders in the fossil fuel industry apparently believed that they were living and doing business chiefly in the world of conventional politics and business. This was a fateful mistake.

In the words of one Canadian observer, "The political Left weaponized the Internet long before the Right even knew what it was."

Canadian oil industry executives (not to mention pipeline executives and Alberta and federal politicians and regulators) discovered this during the so-called pipeline wars, when they were badly mauled by American and Canadian environmentalists who knew how to operate in the Mind Field. Many industry leaders had little understanding of how to win public fights, let alone the dictum, "if you want to be heard, start by listening."

In fairness, public hearings in which one faces opponents who are local, militant, and eloquent aren't a level playing field. Most fossil fuel industry executives were accustomed to using facts, figures, and official language; their adversaries were experienced and effective at using half-truths, symbols, and sensational claims, and they knew how to play headline writers like violins. Historically, understanding and influencing public opinion was never the fossil fuel industry's strength.

An early example of their approach to public engagement occurred in 1928, when the head of the Anglo-Persian Oil Company rented Achnacarry Castle in Lochaber, Scotland, and invited petroleum industry leaders for two weeks of fine dining, grouse shooting, and salmon fishing as well as the hammering out of an unwritten agreement to fix the price of petroleum and steady a volatile market.

The so-called Achnacarry Agreement protected the interests of the Anglo-Persian Oil Company (later BP), Gulf Oil, Shell, and Exxon, and it controlled the world price of oil until the rise of OPEC in the late 1960s. It was an understanding among gentlemen, so it was agreed that the text of the agreement wouldn't be signed—always a good idea when you're creating a cartel to fix prices and control input.[48]

The sometimes cringe-inducing history of Big Oil's dealings with governments and politicians in places like the Middle East, Africa, and Latin America, not to mention Louisiana, has been dealt with elsewhere.[49] Often, the industry advanced its interests with unsavory methods. Corrupt regimes were propped up. Grease was applied. Notoriously, in 1950s Iran, the industry collaborated with the CIA to overthrow the elected Mossadegh government and install the Shah. A generation later, this led to the Islamic Revolution, war, and an enmity with America that has lasted 40 years.

In Canada, the industry's historical conduct was, on the whole, appropriate. Major companies operated respectfully and, over the years, grew executives like Frank Spragins, Eric Newell, and Rick George who had a social conscience and were sensitive to politics and changing social customs. But when the Mind Field underwent a phase change early in the 21st century, the Canadian petroleum industry still had too many technocratic executives who thought they were still living in the 1990s.

As we saw earlier, the initial development of Turner Valley, Alberta's first oil field, was an unregulated gold rush with minimal government control. The result was a debacle for both the industry and Alberta: wild fluctuations in prices, over drilling, and depressurization that wasted natural gas and left far too much oil in the ground, unrecoverable, Starting in 1938, the young Social Credit government resolved to learn from that and began to regulate Turner Valley. By the time oil was discovered at Leduc in 1947, Ernest Manning's administration had put a secure ring of regulations around the industry and a regulatory wall between it and cabinet.

Manning was nervous about the industry's potential to corrupt politics, so he forced it to jump through the hoops of the Oil and Gas Conservation Board before getting any government approvals. He deviated from this practice only once, when he made a side deal with J. Howard Pew in order to kick-start oil sands development. The content of the deal to support Great Canadian Oil Sands was above board; only the process of making it was unusual.

By the late 1960s, the oil and gas industry internalized the rules of engagement as they were then understood and achieved a modus vivendi with the Alberta government. Oilmen of the day had a practical understanding of public opinion and the realities of dealing with the Alberta government. It's true that, in those days, the public opinion environment was infinitely less complicated and more placid than it became once the internet and digital communication revolution came into being. If you wanted to influence mainstream public opinion, you had to get your story covered in the papers, particularly *The Globe and Mail*. If you wanted to ask the government for something, you had to go through channels, document your case, and meet with officials. Under the right circumstances, you could get a meeting with the premier, but before that, you had to dot your I's and cross your T's, and even still, the outcome was far from certain.

In those times, Spragins was a more insightful student of politics than your average oil industry executive. He was instrumental in supporting the development of Syncrude's sophisticated and open communication program, one of the industry's most advanced.[50] However, even he missed signals that the rules were changing. In 1972, after Peter Lougheed was elected premier, and Spragins requested a meeting to discuss the government's intentions toward future oil sand development. At the beginning of the meeting, since they knew each other, Spragins called him "Peter." Lougheed stiffy instructed him to say, "Mr. Lougheed."

Some Things about the Mind Field Many Fossil Fuel Executives Were Slow to Understand

A full analysis of the ways in which fossil fuel industry leaders—and many other North American business leaders—failed to grasp how the Mind Field changed the rules of business would take more space than we have available.

Here are a few of the main points:

- They overrated the importance of facts and logic and underrated the power of emotions and intentions. In the Mind Field, there's no generally accepted body of facts on which everyone agrees. Every advocate arrives at the debate armed with their own facts and version of the truth. Debate is driven by a complex mixture of facts that people select, plus feelings, style, intentions, and even tone of voice.
- Debates are dominated by 15-second sound bytes and oversimplifications. There's no room for subtlety, fine distinctions, or complex arguments. If it sounds like you're going down the road of complexity, you're tuned out.
- Business executives who came to power in the 1980s and 1990s learned the rules of stakeholder relations. Whether the stakeholders were financial analysts, government officials, or representatives of local communities, they knew how to speak to them, how to reason with them. As these executives learned during the pipeline wars around 2010, anti-pipeline campaigners weren't stakeholders. They didn't come to public hearings to discuss the impact of a pipeline; they came to stop it.

- An old RCMP officer who used to teach young cadets liked to say, "In street fighting the first rule is that there are no rules."

- The content of the traditional mainstream media was moderated; newspapers, magazines, film, and television all had editors who were responsible for some level of verification and fact-checking. If you had a problem with something a newspaper published or a TV network broadcasted, there was somebody to complain to. By contrast, the Mind Field is unmoderated: anyone can publish anything, and there's nobody you can complain to about bias or untruthfulness. This was famously satirized in an SCTV skit about a fictitious tabloid called *The National Midnight Star*. The publisher of the paper, portrayed by Eugene Levy, said, "I don't understand why people complain about our stories. We *believe* in verification. Our rule is, *somebody* said it or we wouldn't have printed it."

- People who are effective in the Mind Field work at advocacy all the time. Most business executives try to participate in it off the corner of their desk.

- In the Mind Field, the trial doesn't take place in a courtroom with rules of evidence or a guideline of innocent until proven guilty. Rather, if somebody makes an allegation against you, it rockets around the world and you're halfway to three-quarters convicted before the trial even begins.

- The Mind Field is fluid; what it finds hot and hip changes all the time. It continually spawns new ideas, issues, and fashions. Oil industry leaders like Rick George came to understand that, but most didn't.

- The Mind Field is an uncomfortable place for people who don't like to advocate ideas or participate in debates. During this period, many fossil fuel industry executives didn't really like debate and weren't comfortable with it. Often, they were and are engineers and people of a technical orientation. There were some exceptions to this rule in the Canadian oil industry—Jim Gray, "Smiling Jack" Gallagher of Dome Petroleum, George of Suncor—but they weren't typical. What was typical was an aphorism most often heard within the four walls of The Petroleum Club, namely "debating is getting into a pissing contest with a skunk." So executives spent their time with stakeholders they understood, like shareholders, Wall Street analysts, or business writers.

As numerous social and cultural commentators have acknowledged, the Mind Field—particularly through the growth of social media and the decline and deterioration of the mainstream news media—fostered the cheapening, coarsening, and distortion of public life in the 21st century.

The Echo Chamber

A final word on one of the dominant features of the Mind Field as we entered the 21st century: its tendency, driven by the rising power of social media, to divide society and polarize debate.

Facebook and other social media provide every person and interest group with a microphone to amplify their voice and a hearing aid to tune out contrary views and receive selective feedback. Take a society of people who communicate to others principally through Facebook and receive curated news feeds only from sources they trust (a plethora of echo chambers). Then add in the growth of the outrage industry that manufactures headlines aimed at capturing clicks and eyeballs. What do you get? A dysfunctional discourse.

The table had been set for the explosive growth of the climate change issue and its exploitation by the climate change movement.

12. CLIMATE CHANGE: THE IDEA AND THE MOVEMENT

"Climate change will mean
the end of the Human race."
David
Attenborough

"For every complex problem there is
an answer that is clear, simple, and wrong."
H.L. Mencken

At different times in recent centuries, the Western world has been swept by powerful movements for social, economic, and political reform. These have included movements to stop slavery, emancipate women, prohibit alcohol, outlaw wars of aggression, and bring about nuclear disarmament. All these movements were driven by justified moral passion and a mission to right wrongs. Some of them deeply changed social customs and altered international relationships. Some of them overturned governments and brought down empires. Some of them appeared destined, at least for a time, to permanently change the order of things and reset the world's moral order. Sometimes, that was a mirage.

In August 1928, all the major governments of the world, including Japan, the Western powers, Germany, and Russia, signed the Kellogg-Briande Pact in which they solemnly declared "that they condemn recourse to war for the solution of international controversies, and renounce it, as an instrument of national policy in their relations with one another" and that "the settlement or solution of all disputes or conflicts of whatever nature or of whatever origin they may be, which may arise among them, shall never be sought except by pacific means."

Six years after the signing of the Kellogg-Briande Pact, Japan invaded China, Italy invaded Ethiopia, and we began the slide into the worst war in history.

In the late 20th century, primed by the rapid expansion of the Mind Field, a new, international, and revolutionary idea was born, namely that human-caused greenhouse gas emissions—principally carbon dioxide and methane—were driving catastrophic climate change and that this process had to be stopped by an international effort to restructure global energy production, decarbonize the economy, and bring an end to the use of fossil fuels. This idea was driven by a powerful moral vision: the preservation of the biosphere that sustains humanity. This was the central *idea* of fighting climate change.

Clearly, if this vision were true—or even probable—the Earth's people and governments needed to address the problem with full urgency.[51] And also with full intelligence.

A fundamental question that the principal actors in the climate change movement didn't answer was *how* to decarbonize the world economy. Simply declaring that "fossil fuels must go" wasn't an intelligent answer. Energy systems are vast, complex, and serve many important purposes, like keeping us alive. In the past, energy systems changed over a long time, like multiple generations or centuries. The larger and more complex an energy system is, the deeper its roots in the economy and the longer it takes to replace. To forecast a "climate crisis" without sufficient evidence is one level of irresponsibility. To posit a solution —"decarbonization"—without intelligent analysis compounds the irresponsibility.

It would be hard to exaggerate how revolutionary the idea of human-caused climate change is. Put aside the question of whether we understand the dynamics of the planet's climate well enough to forecast how it changes. Even if we do and we're forecasting it accurately—two very big and unproven assumptions—do we really know how to manage the entire planet's economy in order to halt the global average temperature at less than 2C by 2050?

The idea of catastrophic human-caused climate change (as opposed to so-called natural fluctuations in climate that have taken place over the Earth's lifetime) gave birth to the *climate change movement*, an international network of environmental groups determined to limit the greenhouse gas content of the Earth's atmosphere and create a world powered by zero-carbon renewable energy.

This chapter will discuss the climate change idea, the climate change movement, and the impact of both on the future of Canada's fossil fuels.

* * *

I Initially, many people in the fossil fuel industry underestimated the strength of the climate change idea and the climate change movement. This was unwise. Since the early 1960s, environmentalists had been fighting dozens and dozens of wars against corporations and governments; more often than not, they won. They knew how to operate in the Mind Field and how to exploit its properties to the fullest. How to identify and describe corporate misbehaviour. How to demonize opponents and arouse indignation. How to arouse public fear and play on it. How to raise funds to campaign. How to marshal evidence, how to mobilize public opinion. Most importantly, they understood the ground rules of politics—US politics especially—and which political levers to pull.

In the 1990s, they learned how to build momentum and generate pressure in the arenas of international diplomacy and UN politics, how to drive the adoption of pacts, protocols, and treaties and then leverage them to pressure the policies of national governments.

The vision of the climate change movement and its mastery of political warfare were to have grave consequences for the fossil fuel industry. In the early years of the 21st century, that warfare focused on the Canadian petroleum industry and its drive for markets.

The Climate Crisis Idea

Awareness of atmospheric carbon dioxide and its possible effects on global climate gently bubbled into the international scientific community after 1824, when French physicist Joseph Fourier discovered the greenhouse effect that impeded the Earth's reflection of solar radiation back into space. Without CO2 and the greenhouse effect, the Earth would be encased in ice, would have an average year-round temperature of -18C, and would be unable to support life. You definitely want to have CO2 in the atmosphere. Just not too much!

Various scientists explored the idea over the years, including nuclear physicist Edward Teller in 1957. In 1963, it was the focus of a first international scientific conference. Scientific concern reached into popular politics in President Lyndon Johnson's 1965 State of the Union Address, and from that point forward, it slowly began to grow in public consciousness and influence mainstream politics in the Western world.

In 1979, a UN group called the World Meteorological Association, meeting in the first World Climate Conference, declared that "carbon dioxide plays a fundamental role in determining the temperature of the earth's atmosphere, and it appears plausible that an increased amount of carbon dioxide in the atmosphere can contribute to a gradual warming . . . but the details of the changes are still poorly understood."[52]

Some oil companies followed the discussion. Between 1979–1982, Exxon conducted a research program of climate change and climate modeling, including a research project of equipping their largest supertanker with ocean-testing equipment. Exxon scientists reported to management that human-caused climate change was real and measurable.[53]

In 1985, famed astronomer Carl Sagan testified before Congress about the greenhouse effect on the planet Venus and warned of its likely warming of our planet's

James Hansen giving testimony before the United States Congress in 1988.

atmosphere.[54] The political takeoff point for the climate change issue is generally thought to be the 1987 Congressional testimony of James Hansen, a space scientist at Columbia University who argued that manmade CO2 emissions were changing the Earth's climate. Hansen supported his testimony with his own impressive research. He thus is generally credited with crystallizing public concern that atmospheric CO2 could create a climate crisis.

For a time, Hansen's report had only a limited impact on mainstream politics or government planning, but that was changing. The idea was beginning to snowball.

In 1988, the United Nations created the Intergovernmental Panel on Climate Change (IPCC) to co-ordinate climate research and develop a public database for climate science. From that point forward, research focused not on whether human-caused (or industrially caused) climate change was happening—that it was happening was clear—but how rapidly the climate was changing, how the changes would impact different parts of the planet, and how best to combat it.

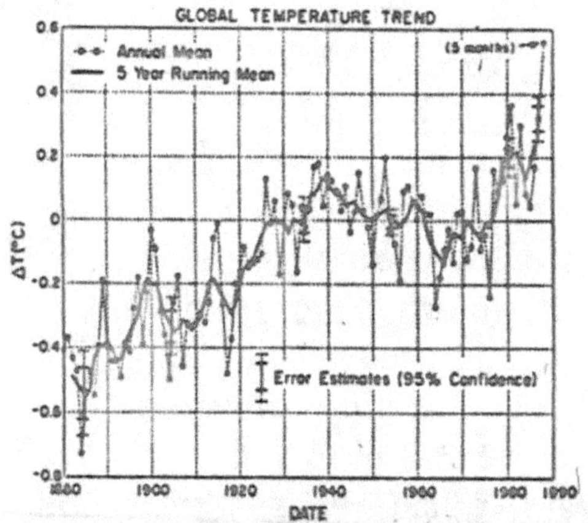

Fig. 1. Global surface air temperature change for the past century, with the zero point defined as the 1951-1980 mean. Uncertainty bars (95% confidence limits) are based on an error analysis as described in reference 6; inner bars refer to the 5-year mean and outer bars to the annual mean. The analyzed uncertainty is a result of incomplete spatial coverage by measurement stations, primarily in ocean areas. The 1988 point compares the January-May 1988 temperature to the mean for the same 5 months in 1951-1980.

James Hansen's global temperature graph. (Figure source: Hansen 1988.)

The differences among what was happening to the climate, why it was happening, and what should and could be done about it were vitally important distinctions for public understanding. Many people—politicians, journalists, policy-makers—didn't get these distinctions or lost sight of them as the ideas percolated through the Mind Field and were converted, distorted, and often weaponized by politicians, activists, scientists, and social and mainstream media into something that was sometimes less about science than about politics or even ideology.

In 2006, former Vice President Al Gore produced *An Inconvenient Truth,* which dramatically summed up the climate change case. The film drew a considerable audience. Public belief in human-caused climate change rose, although at different rates in the United States and Europe. In a July 2007 internet survey of 47 countries conducted by The Nielsen

Al Gore's hearing on global warming. (Photo credit: Wikipedia.)

Company and Oxford University, 66 percent of respondents who'd seen *An Inconvenient Truth* said it had changed their mind about global warming, and 89 percent said it made them more aware of the problem. Three out of four (74 percent) said they changed some of their habits after seeing the film.

Gore announced that he was building an organization to train 1,000 climate change ambassadors—idealistic young people—to carry the message to communities and universities. An international poll indicated the widening split between American and European public opinion on climate change. In 2007, Gore was awarded the Nobel Peace Prize along with the IPCC.

Some US fossil fuel interests, especially coal companies, attacked the central idea that human-caused CO_2 was responsible for climate change or that climate changes were outside normal long-term variability. The weight of science was on the other side. The industry's arguments garnered little support from climate scientists and had little emotional resonance with people at large.

Gore had a considerable political following in the Democratic Party; his promotion of the idea of a climate crisis raised its profile dramatically but also deepened partisan divisions on the issue (i.e., anything *they* propose we must oppose). Fox News and the Republican right attacked Gore as a hypocrite living in an enormous, energy-inefficient house and travelling the world by a greenhouse gas–emitting private jet. Gore and his allies shot back, characterizing anyone who questioned the propositions of *An Inconvenient Truth* as a stooge of the fossil fuel industry, if not worse

Environmental activists began to label anyone who questioned any aspect of climate change theory as a "climate change denier," a manipulative phrase with a not-accidental similarity to "Holocaust denier." Eventually, most of the fossil fuel industry stopped arguing against human-propelled climate change. In 2007, ExxonMobil "recognized that climate change is a serious issue [and] that greenhouse gas emissions are one of the factors affecting climate change" and announced that it would "discontinue contributions" to research groups that question climate change.

Instead of questioning whether human-induced climate change was happening, the industry would have been better to ask whether decarbonization could be achieved in a matter of a few decades or whether major reductions in global emissions could be made before mid-century. That debate never took place until after the climate crisis proposition sunk deep roots.

The Debate That Never Happened

A number of scientists, economists, and energy analysts took issue with different components of the climate crisis proposition. Their critiques covered a wide spectrum and generated responses ranging from thoughtful to vicious. Three of them—Bjorn Lomborg, Steven E. Koonin, and Vaclav Smil—published critical analyses that merited wider awareness.[55]

Three dissenters. *Left*, Lomborg. (Photography by Emil Jupin.) *Centre*, Koonin. (Photo credit: Wikipedia.) *Right*, Smil. (Photo credit: Olibroman at English Wikipedia, licensed by CC BY-SA 3.0.)

Lomborg is a Danish economist specializing in the study of environmental policy. Koonin is a physicist, climate researcher, and former under secretary of energy in the Obama Administration. Smil is a scientist who's published some 40 books and 500 papers on energy and environment and has the distinction of being Bill Gates's favourite writer (and probably the most important thinker few people have ever heard of).

While their works were reported in the business media, social media, and conservative circles, they received scant attention in the mainstream media. They're worth encapsulating here since they have enormous relevance to any discussion on the future of oil and the oil sands. I'd summarize them this way:

- the atmosphere *is undeniably* warming, partly due to industrial and agricultural greenhouse gas emissions, but the *rate* of actual warming has been exaggerated by the climate change movement and comes nowhere near the most alarming predictions;

- many of the extreme predictions of damage from global warming—widescale deaths due to climate events, widescale food shortages from changing weather, and enormous costs to the world economy—have not, to date, come even close to being borne out. For example, deaths from climate events have fallen dramatically over the past century even though the climate has warmed between 1C and 2C. Food yields have improved as the climate has warmed and global economic progress has continued. The data do not support that extreme weather events like hurricanes have increased due to climate warming.

- Long-term predictions of climate disaster shouldn't be based on climate models that vary widely in reliability and generally fail to take into account the complexities of climate dynamics. Predictions based on these models aren't sufficiently reliable to support revolutionary climate policies.

- Promoters of decarbonization have grossly understated the difficulty of achieving it. The means that they propose for doing so—widespread adoption of renewable energy plus suppression of fossil fuel production and consumption—are either *unworkable* (renewables can't be sufficiently scaled up in time to achieve greenhouse gas targets), *undesirable* for a large part of the human race (because they would deny developing countries access to the energy they require to improve living standards), or *politically unworkable* (they would require developed countries to shut down vital industries); and

- policy-makers, for all these reasons, would be well advised to cool the overheated rhetoric of climate change and greatly increase investment in adaptation to it and technological innovation to produce energy with far fewer emissions.

One reviewer summed up Smil's latest findings this way:

> Designing hypothetical roadmaps outlining complete elimination of fossil carbon from the global energy supply by 2050 is nothing but an exercise in wishful thinking that ignores fundamental physical realities. And it is no less unrealistic to propose legislation, as has been done in the U.S. Congress, claiming that such a shift can be accomplished in the U.S. by 2030. Such claims are simply too extreme to be defended as aspirational. The complete decarbonization of the global energy supply will be an extremely challenging undertaking of an unprecedented scale and complexity that will not be accomplished —even in the case of sustained, dedicated and extraordinarily costly commitment—in a matter of a few decades.[56]

None of these critics argue that human-caused climate change isn't happening, only that predictions of catastrophe are unsupported by data and observation and that many of the proposed remedies are impracticable or likely to fail. Having said that, unless you normally read business media, conservative journals or the writings of experts not covered in the mainstream media, you may well find these ideas shocking if not horrifying.

The climate change movement—and to a large extent the mainstream media—has either ignored these ideas, attacked them as a smokescreen for the fossil fuel industry, or ridiculed them as denialism. Yet they're supported by serious analysis, and since they go to the heart of the climate crisis proposition, they merit consideration.

Apocalyptical Thinking, Ecological Pessimism, and the Climate Change Doctrine

The central propositions of a climate *crisis* have gained widespread acceptance more from a hammering by the mainstream media and their amplification in social media than from any widespread study of the science. Why have the forecasts and prescriptions of climate crisis been so widely embraced both in the mainstream media and on social media? I have a theory.

If you've followed popular journalism, film, and literary fiction over the past 40 years, you've maybe noted the steady growth of apocalyptic thinking and end-of-the-world themes. The preoccupation of this outpouring of novels, essays, films, sermons, criticism, television specials, and documentaries is that humanity is living under the threat of catastrophe and extinction whether by climate change, nuclear war, nuclear plant meltdowns, chemical pollution, overpopulation, terrifying pandemics, genetic experimentation, artificial intelligence, or being struck by comet, meteor, or asteroid.

All of these risks are *possible*, and ignoring any of them would be foolish. Failing to assess and analyze them—and subject them to risk assessment—would be equally foolish. But that is what far too many people have done. Instead, they have embraced these catastrophic scenarios. This has contributed to a widespread sense of *existential dread*. Memes of catastrophe have spread through our culture and through the educated classes, and they've been magnified by both conventional and social media.

The persistence of existential dread isn't about "the science." It's about social psychology.

Imaginings of existential threats have become what French syndicalist Georges Sorel called "a myth," one that has been absorbed into the public's unconsciousness. Sorel wasn't equating "myth" with "lie," and neither am I. He wrote that myths "are not banal descriptions of the desired society but calls for action . . . a myth *cannot be refuted*, since it is, at bottom, identical with the convictions of a group." Or as Isaiah Berlin put it, "the function of myth is to create an epic state of mind." Can you imagine a more vivid and epic state of mind than the ending of life on Earth?

The myth of climate crisis is that we're on the brink of catastrophic climate change caused by greenhouse gases, that it can only be averted by the abandonment of fossil fuels, that early abandonment of fossil fuels is possible and can be done without endangering the world economy; that we need to immediately build a new Earth without coal, oil, natural gas, and nuclear plants, a world powered by wind turbines and solar panels, an arcadia in harmony with nature. And not just in a century, but tomorrow (or, at worst, by 2050).

Lomborg, Koonin, and Smil contend that this myth has many flaws. This hasn't stopped many people from believing it. Thanks to existential dread, they *want* to believe in it. And when people want to believe something, facts and logical analysis are swept aside.

The Emergence of the Climate Change Movement

By themselves, ideas don't change politics. Ideas need to be taken up and put to work by political *movements* that convert them into forces pushing for political change. Which brings us to the movement that has converted the idea of a climate crisis into a political force. (By "movement," I don't mean an organized entity with a recognized leader and a formal dogma or doctrine but rather a loose association or network of activists and intellectual influencers who often act independently but share a common set of beliefs and goals. For example, Greenpeace is a cross between a foundation, a non-profit corporation, and a political party. It isn't, in itself, a movement.)[57]

A full analysis of the internal dynamics of movements—the ways in which people exchange ideas, develop plans, and take part in actions—is beyond the scope of this book, so we'll think about the climate change movement in a general way only. What's important to note is that the climate change movement:

- has deep roots in environment groups in Europe and North America;
- is inspired by influential figures on both sides of the Atlantic;
- has strong ties with European green parties and international social movements (e.g., climate justice);
- is politically sophisticated and very experienced at organizing political campaigns (neither of these things can be said about the fossil fuel industry);
- is very well financed by a large network of U.S. foundations and donors as well as by grass-roots subscribers in North America, including Canada;
- has a great many followers and participants in universities, foundations, political parties, and international circles

that often work through the UN as well as within both social media and the mainstream news media.

These conclusions aren't the contentions of some conservative critic. They were documented by a forensic inquiry into the climate change movement in 2020–2021 headed by J. Stephens Allan. The work of the Allan inquiry will be referenced further below.

The War on Fossil Fuels

From the 1960s to the 1990s—roughly the period prior to the rise of the climate change issue—the North American environmental movement chiefly focussed on issues like stopping nuclear power (and, before that, nuclear testing) and reforming or blocking industries, including chemicals, mining, and forest products.

On North America's west coast, the birthplace of the environmental movement, people campaigned to save the redwoods and stop oil exploration in Alaska. In British Columbia, environmental groups like Greenpeace and the West Coast Wilderness Committee targeted the so-called industrial forest industry and campaigned to save old-growth forests and the western spotted owl as well as to stop the use of chlorine in pulp and paper manufacture.

Many of these campaigns succeeded at least

partially. In British Columbia, environmentalists stopped most clear-cutting of old-growth forests and caused government to close vast stretches of the province to development and designate large areas as provincial parks. They also dealt a blow to the province's largest forest company, MacMillan-Bloedel, a blow from which it didn't recover.

Starting in the early 1990s, the concept of a human-caused climate crisis began to be taken up by established environmental organizations like Greenpeace, the Sierra Club, and the Natural Resources Defence Council (in Canada, by the David Suzuki Foundation and others) and drew these groups into a potent international movement, a force in the Mind Field that deeply influenced government energy policy around the world, promoted wind and solar power, and buffeted Western fossil fuel companies, although not the government-owned fossil fuel companies of the Middle East, Russia, and China.

Initially, the movement didn't single out the oil sands industry for special attention. Its target was more broadly US fossil fuels, chiefly offshore oil and gas drilling, the coal industry, and the fracking industry while of course maintaining a steady drumbeat against Big Oil (the movement had already ring-fenced the nuclear energy industry years before).[58] For years, environmentalists demanded change in people's consumption of energy (coal and petroleum especially).

At a certain point after the turn of the century, US environmental strategists realized that this demand was getting little traction: fossil fuels were just so *useful* and energy dense, and so many industries had been built on them.

The movement didn't focus on the Canadian oil sands industry until after 2005, when it coined the "dirty tar sands oil" campaign and the fight to stop oil sands pipelines. Those campaigns will be addressed below.

First, here's an overview of the climate crisis strategies and tactics that were used to shape public opinion in the Mind Field.

Advocacy

One of the demonstrated strengths of the environmental movement has been *advocacy*—the ability not just to manufacture headlines but more broadly to mine scientific findings for data that will support compelling arguments for policy. These data are then amplified by mainstream and social media into trends that sweep up public support.

Most of the data pool for the climate change movement came from reports of the IPCC. The IPCC was created in 1988 by the United Nations Environmental Program. Its stated purpose was "to provide regular assessments of the scientific basis of climate change, its impacts and future risks, and options for adaptation and mitigation." Over the past 30 years, the movement has been very adept at mining the publications of the IPCC for intellectual ammunition to support its calls for change. People deeply familiar with the contents of those reports have pointed out that climate change activists and their supporters in the media have frequently cherry-picked IPCC data or even distorted IPCC findings (report summaries especially) in order to drive the idea that the science behind the IPCC's forecasts of extreme climate change is settled.[59]

Most of us are familiar with consultants' reports that make recommendations and then go up on bookshelves, never to be taken down again; I have written such reports myself. The IPCC's reports weren't used in that way. The climate change movement mined the reports for ammunition, found plenty of it, and translated (and often oversimplified or distorted) its findings into explosive imagery and political rhetoric. This rhetoric deeply influenced Western politics and constituted, as the IPCC itself said, "a key input into international climate change negotiations."

The weaponization of IPCC findings was difficult to prevent because the organization's reports are famously massive and often couched in dense technical language. Like the Bible in an earlier era, everyone alluded to the reports, some people directly quoted them, others paraphrased or slanted them, and, in the media at least, almost no one who cited them had actually read them. They were hard to read; densely larded with climate science jargon and concepts that would be opaque to anyone not trained in statistics. This made it easy for climate change extremists, to cite the IPCC as a source, talk about settled science, and make horrifying prophecies (e.g., that civilization will end in 12 years) without fear of being challenged by the IPCC or anyone in the movement.

The climate change movement (not the IPCC) has been an accomplished fashioner of misleading, manipulative headlines:

> Climate change will lead to the slaughter, death and starvation of six billion people in this century. (Roger Hallam, co-founder of Extinction Rebellion)

> Climate change is the single biggest thing that humans have ever done on this planet. The one thing that needs to be bigger is our movement to stop it. (Bill McKibben)

> It's not that we have a philosophical difference with the fossil fuel industry, it's that their business model is destroying the planet. (Bill McKibben)

In time, climate change activists aimed their

rhetorical guns at the oil sands industry:

> Tar sands oil is the dirtiest fuel on Earth. (Frances Beinecke, former head of Natural Resources Defence Council)

> [The tar sands are] Canada's most shameful environmental secret . . . the largest—and most destructive—industrial project in human history. (Tzeporah Berman, Greenpeace)

> Already millions of barrels of tar sands oil have been extracted from under the Canadian wilderness, producing three to four times more greenhouse gas emissions than conventional oil extraction and using enough natural gas every day to heat three million Canadian homes. Add to this the mass deforestation the project is causing and it becomes clear that *the tar sands must be shut down* if we are serious about tackling disastrous climate change . . . tar sands development has been labelled "the most destructive project on Earth." (UK Tarsands Network, italics added)

> Oil from tar sands is one of the most destructive, carbon-intensive and toxic fuels on the planet. Producing it releases three times as

much greenhouse gas pollution as conventional crude oil does. Tar sands oil comes from a solid mass that must be extracted via energy intensive steam injection or destructive strip mining, techniques that completely destroy ecosystems, put wildlife at risk, and defile large areas of land. Finally, when transported by pipeline or rail, it puts communities, wildlife and water supplies *in danger of toxic spills that are nearly impossible to clean up.* Here's why: *Dirty tar sands and other destructive fossil fuel projects pose a huge risk not only to people and wildlife but to the future of a livable planet.* (Centre for Biological Diversity, italics added)

The ugliest environmental disaster that I not only have ever seen but that I could comprehend, . . . Fort McMurray looks like Hiroshima, not because the houses in Fort Mac stand for 50 sites [but] because the name Fort Mac stands for diseases that these First Nations people are getting, pollution, everything that's happening there. We made a deal with these people. We are breaking our promise. *We are killing these people. The blood of these people will be on modern Canada's hands.* (Neil

Young, musician and songwriter, italics added)

On a spectrum of irresponsible rhetoric ranging from mildly misleading to demonstrably and scandalously false, most of these statements would cluster close to the latter. But that doesn't mean they failed to generate headlines.

National and International Political Action

The movement's strategy for influencing government policy resembles what marketers called push and pull. It works like this: use advocacy campaigns (in the mainstream media, social media, direct mail, etc.), voter mobilization campaigns, high-profile street demonstrations, and intensive lobbying to *push* governments to adopt emissions targets (Kyoto, Copenhagen, Paris, and Glasgow), then use those emission targets to *pull* business and industry into "compliance."

In practice, this enables a national government like Canada to meet with the oil industry and say, "Sorry, everyone. We wish we could accommodate your needs, but our hands are tied by treaties." It also enables a government to tell climate activists that "we're working hard toward fulfilling our commitments," even if that government isn't actually reducing the world's greenhouse gas emissions.

Climate change strategists believed—not without reason—that the soft underbelly of world opinion was international agencies that could be influenced by diplomatic action to embrace treaties, pacts, and covenants, all of which create a sense of unstoppable change and momentum—or at least the impression of them.

When it came to encircling the fossil fuel industry, the climate change movement employed this push and pull strategy. It worked like this: : organize international conferences to create treaties and protocols to set limits on greenhouse gas emissions, then use these commitments to pressure national governments to restrict emissions. Use that in turn to leverage restrictions on fossil fuel production. From the 1990s on, the management of conferences, treaties, and protocols created a more or less continual drumbeat of change that echoed through the media and government circles.

Let's have a look at some of the highlights from these conferences.

- **The Kyoto Protocol** was adopted in 1997 and entered into force in 2005. Ultimately, 192 parties signed the protocol, which pursued the reduction of greenhouse gases —carbon dioxide, methane, nitrous oxide, hydrofluorocarbons, perfluorocarbons, and sulfur hexafluoride—concentrations in the atmosphere to "a level that would prevent dangerous anthropogenic interference with the climate system." In 2005, the Kyoto Protocol took effect but without the United States, Russia, and Canada onboard, so its impact was significantly blunted.

- **The 2009 Copenhagen Accord**, meant to replace the Kyoto Protocol after its expiration, resulted in supportive rhetoric but no commitment to a binding successor to Kyoto. Signees made a moral commitment to limit greenhouse gas emissions enough to keep increases in global temperatures to less than 2C.

- **The 2015 Paris Accord** aimed to substantially reduce global greenhouse gas emissions in an effort to limit this century's global temperature increase to 2C above preindustrial levels while pursuing the means to limit the increase to 1.5C. It was signed by 190 states, including the United States and China. In 2017, Trump announced that his country would cease all participation in the Paris Agreement, contending that it'd undermine the US economy and put the country "at a permanent disadvantage." Once Joe Biden was inaugurated, the Unite States formally rejoined on February 19, 2021.

- **The 2021 Glasgow Accord**, in the words of the UN press release, committed nations to a pact that'd "turn the 2020s into a decade of climate action and support." The conference supporting it was attended by some 40,000 people. According to the United Nations, "Nations strengthened efforts to build resilience to climate change, to curb greenhouse gas emissions and to provide the necessary finance for both. Nations reaffirmed their duty to fulfill the pledge of providing 100 billion dollars annually from developed to developing countries. And they collectively agreed to work to reduce the gap between existing emission reduction plans and what is required to reduce emissions, so that the rise in the global average temperature can be limited to 1.5 degrees. For the first time, nations were called upon to phase down unabated coal power and inefficient subsidies for fossil fuels."[60]

How much have these conferences, accords, treaties, and protocols changed the world's energy production and consumption? To what extent have they reduced overall emissions? Have they stabilized or even reduced the buildup of greenhouse gases in the atmosphere? These are questions now asked even by some climate change writers and observers.

Astronomers talk about the difficulty of separating a "signal" from the "noise" generated by the cosmos. In the same way it's hard to separate climate crisis noise (headlines, street theatre, meaningless proclamations, manifestoes, Panglossian declarations by politicians, etc.) from binding commitments and actual government policies with teeth and resources behind them and to separate *that* from actual climate data.

The UN's final report on the Glasgow Climate Pact said,

> Greenhouse-gas emissions must be reduced and carbon dioxide emissions must fall by 45 percent from 2010 levels by 2030 for global warming to be maintained at 1.5C above pre-industrial levels. . . . *[U]nder existing emissions-reduction pledges, emissions will be nearly 14 percent higher by 2030 than in 2010.* Countries acknowledged the need to reduce emissions faster. (Italics added)

Following Glasgow, Niklas Hoehne, a Dutch climate researcher, wrote that countries need to make more ambitious pledges to tackle climate change. "COP26 has closed the gap, but it has not solved the problem," he told *Nature* magazine. "Modelling suggests that the promises will still not be enough to limit global warming to 2C above pre-industrial levels, the goal stated in the 2015 Paris climate agreement. [Even] if countries meet their 2030 targets, global temperatures will still rise 2.4C above pre-industrial levels by 2100."[61]

Many of the pledges to reduce emissions have not been fulfilled. Where emissions reductions have taken place, they have often been counteracted by increased emissions in other places. At Glasgow, commitments to drastically reduce coal consumption were watered down into a pledge to phase it down, after which China and India, two of the world's leading coal producers and consumers, said they had no intention of doing either.

In short, there's evidence to suggest that all the *sturm und drang* of international climate conferences has not brought us even close to the goal of net zero by 2050.

What if all these treaties, pacts, covenants, and commitments don't change fossil fuel production and consumption enough to come anywhere near meeting targets? What happens if the movement finds that it can win the war in the Mind Field and lose it in the economy?[62]

Demonstrations and Street Theatre

One of the major strengths of the climate change movement that grew out of the environmental movement's experience before it was its ability to capture public attention by staging publicity-rich street demonstrations and mass protests (not to mention online petitions, open letters to government, UN resolutions, etc.). Its tactics provided the news media with a rich diet of easily packaged photo ops and sent a message to politicians on both sides of the Atlantic that the demand for climate change policies was real, visible, and to be respected.

This was just another example of how the climate change movement realized how to play to its strengths and the fossil fuel industry's weaknesses. All of these tactics were things the climate change movement did well and the fossil fuel industry did poorly or not at all.

In the 2000s, large-scale demonstrations calling for action on climate change became regular events. Bill McKibben's 350.org network connected with hundreds of grassroots organizations to organize global days of action in dozens of countries.

Extinction Rebellion, launched in London in May 2018, soon elbowed its way into the European media with so-called days of rage that mobilized tens of thousands of people, many of them adolescents, to block bridges, tie up traffic, and confront anyone who dared to argue with them. The group brought the U.K. capital to a standstill and began targeting events like Fashion Week in order to further their mission. Chapters of the group subsequently opened in 68 countries, from Russia to South Africa, although its actions in authoritarian states like Russia were largely invisible. Climate activism was on the march.

These protests claimed to be the largest climate strikes in world history; organizers claimed that over 4 million people participated in them in the United Kingdom, Germany, Australia, and many other countries. In New York—where Greta Thunberg delivered a speech —some 250,000 people turned out on the streets. In country after country, it rolled on.

September 2019 climate strike in Geneva. (MHM55, Own work, licensed by CC BY-SA 4.0.)

Then 15-year-old Greta Thunberg went on her first school strike, sitting alone outside the Swedish parliament. Thunberg's passionate, tearful speeches helped her become a media icon; she made the cover of *Time*, and she was even nominated for a Nobel Peace Prize, (the Nobel committee members declined).

The demonstration pressure was soon supplemented by the transition of the climate change movement into schools in Europe and North America. This took two forms: pressure on teachers to reference climate change with their students. This call, which wasn't limited to older students but included teaching even in primary schools and kindergartens, so increased "climate change anxiety" among so many children that the Royal College of Psychiatrists felt impelled to express its concerns about the impact on children's mental health. The College did not call for an end to teachers frightening their pupils, simply increased "dialogue."[63]

Other Creative Tactics: Shareholder Activism and Disinvestment

After 2015, partly as a consequence of the anti–oil sands campaign by climate change campaigners, which included demands that investors disinvest from the oil sands and other sources of "dirty oil", and partly because of the questionable economics of new, higher-cost oil sands projects (older projects that had repaid some or all of their original investment were safer), several European-based oil companies announced that they were selling off or writing down their oil sands assets. Devon Energy, Total Oil and Gas, Royal Dutch Shell, Equinor, and Conoco Phillips all moved in this direction.

Total even wrote off its $9.3 billion in oil sands assets and cancelled its membership in the Canadian Association of Petroleum Producers. In a statement, the company said, "[We] now consider oil reserves with high production costs that are to be produced more than 20 years in the future to be 'stranded' given our carbon reduction targets and because the resource may not be produced by 2050."

Whether it has damaged the industry's ability to finance new investment is questionable since the companies have considerable financial strength and the ability to fund new investment out of their earnings. An ironic consequence of the divestment was the increasing Canadianization of the oil sands industry. The Big Five oil sands companies now have a substantially larger share of Canadian ownership than ever before.

"Cracking" corporate boards has been slow going. In May 2021, Engine Number One, a small hedge fund, campaigned for the election of reform-oriented candidates to the board of ExxonMobil and succeeded in getting three of them elected over the opposition of the company's executive.[64] The same year, a Dutch court ruled that Royal Dutch Shell must reduce the 2030 CO2 emissions of its 1,100 companies by 45 percent from 2019 levels. This is being appealed. It is questionable whether the courts of other countries will follow suit but the future impact of legal actions is speculative.

Sometime between 2005 and 2008 —participants' accounts differ—the climate change movement began a campaign to shut down the oil sands industry. How it was planned, executed, and financed is outlined below.

The Campaign to Ring-Fence the Canadian Petroleum Industry

J. Stephens Allan, Energy Commissioner. (Photo credit: Alberta Trades Hall of Fame Alberta.ca.)

In 2019, the Alberta government appointed forensic accountant J. Stephens Allan of Deloitte LLP as a commissioner to "inquire into anti-Alberta energy campaigns that are supported, in whole or in part, by foreign organizations." Allan's 650-page report was tabled in July 2021.

The government asked Allan:

> *(a)   whether any foreign organization that has evinced an intent harmful or injurious to the Alberta oil and gas industry has provided financial assistance to a Canadian organization that has disseminated misleading or false information about the Alberta oil and gas industry.*

Allan's answer: Between 2003 and 2019, $925 million was donated by foreign donors [many of them US foundations] to Canadian charities. And there's evidence to suggest that much of this money went to Canadian environmental organizations for so-called environmental initiatives. The largest single environmental initiative during this period was the fight to discredit, encircle, and stop the oil sands, so it's reasonable to assume that a large part of this funding went to the anti–oil sands campaigns.

Allan added, "I was ultimately not able to trace with precision the quantum of foreign funding applied to the anti–Alberta energy campaigns . . . due to the fungible nature of money. Once funds are deployed to an organization in some manner, they are deployed to advance the mission and campaigns of the organization, which are often varied and complex, and cannot be readily traced to any particular activity or initiative." Allan was forensic accountant familiar with the difficulty of tracing laundered money.

Allan found that foreign funders had provided $352 million for environmental initiatives that remained in the United States (e.g., such as the campaign to stop so-called Canadian pipeline proposals like Keystone XL).

(b)   whether any Canadian organization referred to in clause (a) has also received grants or other discretionary funding from the government of Alberta, from municipal, provincial or territorial governments in Canada or from the Government of Canada

Allan's answer: During the same period, Canadian governments paid $145 million to Canadian charities for environmental campaigns. Almost all of these campaigns were run by tax-exempt charities. He did not address how many of these campaigns were part of the campaign to stop the oil sands industry.

The Allan inquiry's forensic accountants searched the public record and online information sources as well as United States and Canadian tax records, and Allan invited environmental organizations and foundations to participate in its inquiry. Prior to the Allan inquiry, information on these matters had been published by independent researchers and talked about here and there, but it had been largely ignored or attacked by environmental groups, which were apparently sensitive about the size of their budgets and the amount of money they'd received from major US donors and foundations. The NDP, which historically campaigned about the evils of foreign capital and foreign ownership of the oil industry, was strangely silent on the role of American money in funding the anti-oil sands campaign.

Few environmental groups—a notable exception being the Pembina Institute—participated in the Allan inquiry. The rest were silent or sent up barrage balloons labelled "free speech" and "political bias." Allan and the forensic accountants went to work investigating paper trails and environmental source documents.

Allan's work was prompted by reports, years earlier, of large sums of US foundation money pouring into Canadian environmental causes. For example, in November 2013, Vivian Krause, an independent researcher on the west coast, reported that within a few months of that year, the Tides USA Foundation donated $3.2 million to anti–oil sands campaigns, including LeadNow, Idle No More, Indigenous Tar Sands, Tanker Free Coast, Pipe Up, Tar Sands Reality Check, the Canadian Youth Climate Coalition, PowerShift, and Save the Salish Sea. Seven donations went to promoting Indigenous solidarity with First Nations. All of these initiatives openly proclaimed their opposition to oil sands pipelines designed to get oil sands bitumen to global markets.[65]

Seven payments mentioned building relationships with First Nations, Indigenous solidarity, and resistance and opposition along pipeline routes. For example, through the Tides Canada Foundation Exchange Fund, Tides USA paid $35,000 to West Coast Environmental Law "to provide legal strategies and communication support to First Nations to constrain tar sands development." Through the Tides Canada Foundation Exchange Fund, Tides USA also paid

$15,000 to the Sierra Club of BC for a project called, Our Coast, Our Call: Mobilizing and Strengthening Opposition to Tanker Expansion on the B.C. Coast. And so it went.

Unsurprisingly, Allan and the forensic accountants at Deloitte found that:

- very large sums of money—something in the neighbourhood of $1.28 billion between 2003 and 2019—had been donated by US foundations and wealthy donors to Canadian-based environmental initiatives;
- some part of that money—Allan was unable to determine exactly how much, thanks to the success of the movement in avoidance of trackable record-keeping —was used to finance a very large, multi-year campaign against the oil sands industry and that a significant part of that campaign was misleading and false;
- what part of this went to finance anti–oil sands and anti-pipeline campaigns couldn't be established since neither the US or Canadian governments tracked such data. But given the scope of such campaigns, it was certainly enormous.

An important finding of the Allan inquiry was scope, planning, and co-ordination of the campaigns.

How It Began

There are different views of how what came to be called the Tar Sands Campaign began. One story is that Alberta environmentalists, chiefly the Pembina Institute, played a role. In its testimony to the Allan inquiry, the Pembina Institute said that, starting in the 1990s, it had been alarmed at the growth of the industry and its lack of attention to environmental issues and that it had appealed to US environmental groups to assist its efforts to publicize the industry's failings. Another story is that an Alberta government decision to put a giant 180-tonne oil sands truck on the National Mall in Washington during the 2006 Smithsonian Folklife Festival alarmed US environmentalists and got them thinking about the implications of expanded Canadian oil production invading America. At the Allan inquiry, the Natural Defence Council's Canadian head said, "It was a pivotal moment. When you bring a tar sands dump truck into our own back yard . . . it was a moment of it sort of being 'They've brought this fight to us.'"

However the environmentalists' campaign came together, there's no doubt that parking a gargantuan truck in the nerve centre of American political activism was one of the biggest political blunders of recent times. Whoever thought of it should be known as the Boy Who Kicked over the Hornet's Nest.

By 2008, the leaders of the American environmental movement had come together to plan the Tar Sands Campaign and had mobilized their considerable planning and co-ordination resources behind it. It was destined to be one of the most comprehensive and heavily financed environmental campaigns ever waged, certainly in North America: a massive, multi-year, multi-pronged assault that included grassroots mobilization, the development of documentaries and research papers, direct action and citizen engagement, exploitation of the land conservation and Indigenous rights issues, and divestment campaigns aimed at banks and investors. It would include a consumer advertising campaign to prevent tourists from visiting Alberta, a UK Tar Sands network to encourage actions on the part of individuals, financial institutions, and EU governments to block tar sands development, a Tar Sands Solutions Network, and land conservation measures to bring First Nations and environmental groups together to block the construction of pipelines and refineries. Research papers to support these initiatives were also included, as was litigation and political activism.

A full and detailed review of the campaign strategy and implementation is contained in parts 2, 3, and 4 of the Allan Report.[66]

That was the plan that was executed.

Tar Sands Campaign Strategy. (Figure source: Corporate Ethnics International, http://www.offsettingresistance.ca/ TarSandsCoalition- StrategyPaper2008.pdf.)

When the campaigns were over, the oil sands industry's goals for expanded markets pretty much lay in ruins. The industry's leaders—and their supporters in the Alberta and federal governments—failed to size up the market they intended to penetrate and were reminiscent of the man who brought a knife to a gunfight.

Attacking the Oil Sands' Connective Tissue

The easiest way to attack the oil sands industry was to attack its connective tissue: the pipelines that it proposed to build to get oil sands petroleum to world markets. In short, to landlock the oil sands.

In June 2011, Hansen, first promulgator of the modern theory that CO2 caused climate change, circulated a white paper to climate scientists. It was entitled "Silence is Deadly," and it concerned the proposed Keystone XL pipeline, which was envisioned to carry oil sands bitumen through the United States.[67] Hansen's thesis was that if enough of the world's coal-fired power plants were closed, the world's climate might be stabilized, but "if the tar sands are thrown into the mix, it is essentially game over." He called on climate scientists to oppose the pipeline and object to the State Department, which had to approve it. McKibben of 350.org picked up the challenge. As he put it in an open letter to environmentalists:

> To call this project a horror is serious understatement. The tar sands have wrecked huge parts of Alberta, disrupting ways of life in Indigenous communities . . . the Keystone pipeline would be a 15-hundred-mile fuse to the biggest carbon bomb on the continent, a way to make it easier and faster to trigger the final overheating of our planet.

McKibben called for mass protests in Washington. That August, hundreds of people gathered outside the White House, intentionally trespassed, and were arrested. The Obama State Department didn't reject the pipeline but damned it with faint praise and sent it into the regulatory process to die a death by a thousand small cuts.

Bill McKibben. (Photo credit: Gage Skidmore, licensed by CC BY-SA 2.0.)

It'd be hard to imagine a more extraordinary example of "thinking big" than a call to transform the Earth's entire energy economy, but around 1990, McKibben launched a crusade to do just that. He arguably became the principal advocate of the climate change idea and the most effective translator of it into a global political movement. Hansen was a scientist; McKibben was an advocate, strategist, and networker *par excellence*. He was very good at three things the fossil fuel industry was dismal at: advocacy, networking, and political strategy.

McKibben, whose 1989 book *The End of Nature* is regarded as the first book for a general audience about climate change, was a formidable behind-the-scenes organizer. He authored a dozen books in 24 languages, spoke and wrote in many countries, and created 350.org, which described itself as "the first planet-wide, grassroots climate change movement." By its own account, 350.org "organized 20,000 rallies around the world in every country save North Korea, spearheaded the resistance to the Keystone Pipeline, and launched the fast-growing fossil fuel divestment movement."

With books, speeches, and limitless networking, McKibben raised steadily increasing alarms about climate change and predicted what would happen to the planet if carbon emissions weren't dramatically reduced. To make the point, the cover of his first book, *The End of Nature*, pictured the entire planet aflame. By proclaiming a global climate crisis that'd need to be addressed with unprecedented global action—not just by the fossil fuel industry but by industry and virtually all the world's governments—McKibben and others dramatically expanded the environmentalist vision.

Piecemeal reforms to industrial practices, McKibben and others argued, would no longer be enough; if the planet was to be saved, then the entire industrialized world would need to make breathtakingly radical changes in its production and use of energy, beginning with the burning of carbon and hydrocarbons. What was at stake no less, was the survival of . . . well, everything: entire species, including humanity itself.

Carbon-emitting industries would have to go, beginning of course with coal, then quickly including fracked oil, and then conventional oil and natural gas. If entire sectors of the economy and the industries, human beings, and communities dependent on them had to undergo wrenching change if not abandonment, then, well, sorry about that. Human beings, having changed the atmosphere with industrialization, would now have to change it back again with a global switch to renewable energy, driven by government intervention and global planning. That was McKibben's proposition.

A major summit of environmental groups in New York in 2008 decided to focus the movement's actions on the Alberta oil sands. The summit was attended by McKibben, representing 350.org, and representatives of most of America's leading environmental groups, including Corporate Ethics International, the Rockefeller Brothers Foundation, the National Resources Defence Council, and Forest Ethics. McKibben wrote:

> Even if it flowed from the ground as sweetly as Saudi crude, the real problem is not the extra carbon [in oil sands petroleum] compared to other oil. It's the sheer amount of oil up there. The second largest pool of it on the planet, of which something like 3% has been used so far. They want to greatly increase the rate of flow. We're going from drinking straws to fire hoses with Keystone XL, so that will send the rate of carbon into the atmosphere soaring.

McGibben was stating what became one of the central myths of the campaign, namely that fencing off, restricting, or ultimately abolishing production from the oil sands—or at least blocking the industry's access to the US Gulf Coast and Asia—would reduce greenhouse gas emissions to the atmosphere and represent a huge first step toward drastically reducing world oil production. This myth was built on a fundamental fallacy about how the modern oil economy works. In the real world, if you choke off the flow of oil sands bitumen to the Gulf

Coast or Asia, those markets simply reach out to much larger oil producers like the Middle East, Russia, and other countries—the so-called sovereign producers—to step in to supply their oil. The sovereign producers have plenty of spare capacity and no significant interest in reducing emissions. In the end, the net reduction of greenhouse gasses to the Earth's atmosphere is precisely zero. This inconvenient fact has never been addressed by climate change activists.

What the climate change movement needed was a target that couldn't use the defensive strategies the domestic US oil and coal industries had used. The oil sands industry and its partner, the pipeline companies, met that bill. For one thing, the oil sands industry and the principal pipeline company could be characterized as *foreign* (Canadian) as well as *an environmental threat*, so activists wouldn't be attacking an American industry. For another, oil sands bitumen had a higher greenhouse gas coefficient than, well, American petroleum or even petroleum from the Middle East (how much higher was debatable, but that was a detail). Most importantly, if the oil sands industry were to grow, it would need to get its oil to markets like the US Gulf Coast and Asia. That meant the pipelines would need to negotiate a complex political approval process with dozens of legally defined points for interception. The political and regulatory systems could be turned into a swamp from which the pipeline industry might not extricate itself.[1]

And so the climate change movement targeted the oil sands industry and its proposed pipeline expansion much in the same way that a wolfpack, pursuing a bison herd, will identify a single vulnerable animal, cut it out from the herd, and eventually bring it down.

American environmentalists weren't long in shifting their focus from the fossil fuel industry in general to the Alberta oil sands and its proposed pipelines. As in the First World War, their attack began with an artillery barrage, this time of inflamed rhetoric, widely echoed in traditional and social media:

American environmentalists weren't long in shifting their focus from the fossil fuel industry in general to the Alberta oil sands and its proposed pipelines. As in the First World War, their attack began with an artillery barrage, this time of inflamed rhetoric, widely echoed in traditional and social media:

- [The tar sands is] Canada's most shameful environmental secret . . . the largest – and most destructive—industrial project in human history. (Tzeporah Berman, Greenpeace)

- Already millions of barrels of tar sands oil have been extracted from under the Canadian wilderness, producing three to four time more greenhouse gas emissions than conventional oil extraction and using enough natural gas every day to heat three million Canadian homes. Add to this the mass deforestation the project is causing and it becomes clear that the tar sands must be shut down if we are serious about tackling disastrous climate change. . . . tar sands development has been labelled "the most destructive project on Earth." (UK Tarsands Network)

- Oil from tar sands is one of the most destructive, carbon-intensive and toxic fuels on the planet. Producing it releases three times as much greenhouse gas pollution as conventional crude oil does. Tar sands oil comes from a solid mass that must be extracted via energy intensive steam injection or destructive strip mining, techniques that completely destroy ecosystems, put wildlife at risk, and defile large areas of land. Finally, when transported by pipeline or rail, it puts communities, wildlife and water supplies in danger of toxic spills that are nearly impossible to clean up. Here's why: Dirty tar sands and other destructive fossil fuel projects pose a huge risk not only to people and wildlife but to the future of a livable planet. (Centre for

> Biological Diversity)
>
> - The ugliest environmental disaster that I not only have ever seen but that I could comprehend, . . . Fort McMurray looks like Hiroshima, not because the houses in Fort Mac stand for 50 sites [but] because the name Fort Mac stands for diseases that these First Nations people are getting, pollution, everything that's happening there. We made a deal with these people. We are breaking our promise. We are killing these people. The blood of these people will be on modern Canada's hands. (Neil Young, musician and songwriter).

Looking back on the period, recognizing that the anti–oil sands thinkers decided to focus on pipelines—the industry's connective tissue—we may wonder why the industry placed such importance on pipelines in the first place.

By the first years of the century, because of the rapid expansion of the oil sands industry, Alberta was producing a large amount of petroleum that needed new markets—existing markets in Canada and the northern United States were saturated—either in Canada, on the US Gulf Coast, or in Asia. A significant and growing part of oil sands production was bitumen, a very low-grade form of petroleum that sold at a 20 percent discount in the

Canadian and northern US refinery markets. There were potential markets for bitumen on the US Gulf Coast and in Asia, both of which had refineries designed to process crude oils like bitumen and who would pay a higher price for it if the bitumen could be delivered to them. That meant new pipelines or extended pipelines from the oil sands and new marine shipping terminals to transport it.

Government go-aheads for pipelines and shipping terminals would require dozens if not hundreds of regulatory approvals by Canadian and US federal and state/provincial agencies, not to mention Canadian and US Indigenous peoples.

Why didn't bitumen producers simply build upgraders, as the oil sands mining companies had? Two reasons, technical and economic. Oil sands mining operations were relatively compact and able to integrate utility plants to make steam and electricity, thus minimizing production cost and conserving energy. In situ operations were smaller and widely dispersed through the forest, too far apart to be serviced by central utilities. While it was possible to build stand-alone upgraders (and Alberta had several), they were inordinately expensive to make and operate. In the words of business journalist Robert Bott, "Most if not all of the huge process vessels are made in Korea, and they are too big to be shipped intact by rail or truck, so they have to be disassembled and reassembled in high-labour-cost North America. That's why most coking capacity is in places with tidewater access by ship or barge—mostly Gulf Coast but also potentially Great Lakes or Mississippi River."[68]

At any rate, even if the industry had built upgraders and converted more bitumen to synthetic crude, it would still have needed new pipelines to reach new markets. So industry and the federal and Alberta governments of the day agreed to minimize costs and maximize revenue by shipping the bitumen to Gulf Coast and Asian refiners that had the capacity to refine it and would be willing to buy it at a better price than it would otherwise fetch. At the time, both the industry and the Albertan and Canadian governments apparently thought it was a common-sense business strategy.

A few oil sands industry veterans (including Brent Scott, then retired) thought that producing and exporting bitumen was a bad idea, but the policy had enough support from the industry and government—and the pipeline companies— that it carried the day from 1990 until 2010. As events demonstrated, it was a perilous policy that put the industry directly into the crosshairs of the climate change movement.

The Gulf Coast could only be reached by a massive new pipeline, Keystone XL. Transportation by rail, a possible option, was expensive and capacity-limited, and it raised visibility and public safety questions. Asian markets, meanwhile, could only be reached with massive new pipelines like the Northern Gateway and TransMountain. Reaching Quebec and the Maritimes required a new line called Energy East.

All of these new pipelines required dozens of regulatory and political approvals from federal and state agencies as well as local American communities, not to mention British Columbian First Nations looking for leverage and recognition and the Government of Quebec, which was increasingly green.

Most importantly, gaining public approval required the industry to master new rules of business arising from revolutionary changes in the world of public opinion formation. Whether you're a corporation, an environmental group, or a political leader, acceptance of your word —in other words public trust—is influenced by a variety of factors, including people's belief systems and whether they trust the messenger. Trust arises out of familiarity, competence, and authenticity. You have to invest in building it and give it time. You can't start asking for public trust the second you step off the train.

In the pipeline wars in states like Nebraska —or, 50 years earlier, in Canada's Northwest Territories when the industry was trying to get Indigenous people to support an Arctic Gas pipeline—the pipeliners needed to appreciate that they were strangers in town, asking local people to support change in their lives. That's why public affairs people used to say, "if you want to be heard, start by listening."

When local environmentalists alleged that Keystone XL would be carrying "dirty oil"—not just petroleum but *corrosive, toxic* oil that would inevitably spill and pollute the water supply of a dried-out region—the American public was predisposed to believe the charge. Wait, they were told, there's more! The industry that produced it, the oil sands, was *already* endangering the Earth's climate worse even than the coal or conventional oil industries.

The industry proposed four pipelines to pump oil sands bitumen to new markets. **Keystone XL** was intended to reach refineries on the US Gulf Coast. **Northern Gateway** was to cross British Columbia to Kitimat and get bitumen into tankers for delivery to refineries in Asia. **Energy East** was to pump bitumen to refineries in Quebec and the Maritimes. And **TransMountain** was to twin an existing pipeline through central British Columbia and deliver bitumen to the Port of Vancouver also for shipment to Asia.

What happened to the pipeline proposals was a debacle. Keystone XL was the target of a very well-orchestrated campaign of opposition in the United States, directed by McKibben and joined by a variety of traditional and ad hoc groups of ranchers, Native Americans, and others across the Midwest. It see-sawed in US politics for 10 years. It was subjected to death by a thousand small cuts by the Obama administration, promoted by the Trump administration, and then finally killed by the Biden administration.

The "dirty oil" story was, at best, a wilful and misleading exaggeration; at worst, it was a medium-sized lie.[69] But that didn't prevent it from generating headlines and attracting followers. S Canadian conservatives developed a counterargument to dirty oil; its theme was that Canadian produced "ethical oil" in a democratic system with good environmental regulation and respect human rights, unlike places in the Middle East, which were corrupt, dictatorial, and had few environmental regulations and little regard for human rights. Even in Canada, this argument gained little traction. Polling by the Canadian Association of Petroleum Producers showed that Canadians *expected* Canadian oil companies to act responsibly, socially, and environmentally; they were Canadian after all. In any case, Canadians told the pollsters, they wanted to support *both* a healthy oil industry and a safe global climate and saw no conflict between the two.

Industry Counterattack

In 2009, the Canadian Association of Oil Producers hired pollsters and communication experts to advise them on countering the environmental campaigns. The association spent significantly, though at least an order of magnitude less than the environmentalists, and it conducted cross-country dialogues on the oil sands to engage the public.

The campaign received some attention in the mainstream media but not nearly enough, and it was too late to effectively counter the avalanche of misinformation on social media. One of the things the industry discovered was that the public engagement process isn't a level playing field; industry is held to a much higher standard of fact or truth than its adversaries. Some observers felt that the industry's effectiveness was also greatly eroded by the 2008 financial crisis, which lowered the public's trust in business in general.

The significant reason that the pipeline applications failed in the United States was that the proponents were asking for public trust. In places like Nebraska—one of the chief battlegrounds—neither state regulators nor residents had any reason to trust a stranger, which is what the pipeline applicant was. Keystone XL was foreign (Canadian) and part of an industry responsible for pipeline breaks that polluted US rivers and soil. As well, perhaps, it was part of the big business system that a few years earlier, in 2008, had managed to destroy the savings and hopes of millions of ordinary Americans.

In Canada, governments blocked the three

pipeline applicants. Energy East was unilaterally and without due process blocked by the premier of Quebec, a move supported by green and anti-pipeline opinion there. Northern Gateway was ruled out by the Trudeau government before it could even get out of the starting gate when the government banned tanker traffic on the British Columbia coast.

At this writing, TransMountain is fighting a war of attrition with a small number of First Nations in British Columbia's interior and a deeply entrenched environmentalist opposition in the Lower Mainland that views the prospect of oil tankers in BC waters as an apocalypse. To quiet political critics, the Trudeau government purchased the TransMountain. At the time of this writing (February 2022), the pipeline is still alive but surrounded with life-support monitors in the political ICU. Efforts are underway to sell it to First Nations groups, and these may come to fruition; if so, it could change the political calculus enough to overcome the resistance of First Nations dissenters (if not green opinion on the coast), but this seems a long-shot bet.

Round one of the pipeline wars went to the climate change movement. The oil sands industry and the pipeline companies barely managed to stay on their feet until the bell.

The Rest of the Fight and How It'll End

Prime Minister Justin Trudeau, a centre-left

politician who probably, deep in his heart, belongs to the Green Party, followed a long established Canadian political tradition and straddled the issue of the oil sands and climate change by trying to ride two horses at once. As this was written, he was still managing to stay atop both horses, but with greater and greater difficulty.

The two-horses tradition has old roots in Canadian politics. During the Second World War, William L. Mackenzie King, Canada's longest-serving prime minister, used it to manage the controversial issue of military conscription, which was bitterly opposed in Quebec and supported everywhere else. Going into the 1944 election, King said Ottawa would pursue a policy that he called "conscription if necessary but not necessarily conscription." After the election, he imposed conscription—but only for domestic duty (i.e., guarding coastal radar stations watching for Japanese submarines).

In the same vein, Trudeau said Ottawa would support both reduced emissions and the oil sands industry. His equestrian strategy didn't fully satisfy environmentalists, Alberta, or the industry, but it perhaps kept an uneasy peace, at least for a time. In politics, that's sometimes as good as it gets.

The climate change movement has sunk deep roots in the mainstream media, social media, and the perceptions of the political class in

Europe and North America. Public opinion has been largely won over to the main propositions of climate change activists, although how long this will remain the case when and if climate change policies lead to skyrocketing energy prices and taxes is something yet to be tested.

If climate change policy comes into obvious conflict with smart energy policy, which one will yield? The war in Ukraine is prompting a wide scale re-thinking of energy policy in Europe and some observers wonder how long it will take for Europe to review its climate change goals. . Some erosion of public support for ambitious climate change goals is beginning to play out as these words were written. Already coal is enjoying a comeback – coal! – and Britain is reviewing its moratoria on fracking and North Sea oil and gas production,

Whether and when climate change policies and regulations will succeed in transforming global energy infrastructure and markets is still playing out. Meanwhile, the oil sands industry contemplates its possible futures.

13. INTO THE FUTURE: THE OIL SANDS AND CANADA'S FOSSIL FUEL INDUSTRY

"The notion of a fast track to a wholesale energy transition runs up against major obstacles—the sheer scale of the energy system that supports the world economy, the need for reliability, the demand for mineral resources for renewables . . . on top of all that is the high cost of a fast transition and the question of who pays for it, especially given the staggering amounts of debt that governments took on in 2020 to fight the health and economic consequences of the coronavirus. . . . for the next few decades, the world's energy supplies will come from a mixed system, one of rivalry and competition among energy choices. In this system, oil will maintain a preeminent position as a global commodity, still the primary fuel that makes the world go round. Some will simply not want to hear that. But it is based on the reality of all the investment already made, lead times for new investment and innovation, supply chains, its central role in transportation, the need for plastics from building blocks of the modern world to hospital operating rooms. . . . as a result, oil —along with natural gas, which now is also a global commodity—will not only continue to play a large role in the world economy, but will also be central in the debates over the environment and climate, and certainly in the strategies of nations and in the contention among them." [70]

> Daniel Yergin,
> Pulitzer Prize–
> winning energy
> journalist

It's early 2022. Threats and prophecies abound. We are said to be in the early stages of an energy transition to a net-zero carbon world. Others say the transition will take much longer —and involve much more pain, turmoil, and even upheaval.

It's the age of credulity and the age of disbelief. At the same time.

Given the state of the world and the inroads that the climate change idea has made into public opinion and policy, what are the prospects for Canada's energy? Starting with the oil sands but also including all the other energy with which Canada is endowed?

Hundreds of thousands of Canadians—people employed directly or indirectly by the fossil fuel industry, people dependent on its tax and other revenues, and the millions of people whose heating, transportation, food, and electricity depend on fossil fuels—have an immediate and large stake in the answer to this question.

How Canadians Think

While the oil and gas industry must feel somewhat under siege by hostile government policy trends, Canadians themselves—real voters and real energy consumers—are ambivalent about so-called climate action.

Yes, opinion polls can be misleading. Public responses are coloured by how questions are asked, how much people actually know about the subject, and, most importantly, the degree to which most of us entertain conflicting ideas at the same time. So Canadian public opinion on energy is as complex (and often muddled) as it is on many subjects. Even so, polls can generate some insights.

One pro-climate change organization said, "Most people can't name a climate policy without prompting. They're not thinking about gas furnaces or even gasoline cars. When they think about actions to address climate change, they're thinking about recycling, drinking water, and maybe plastic straws or tree-planting."[71]

In 2021, in a study sponsored by the Canada Action Coalition (a pro-oil and gas organization), three of four Canadian participants agreed the country should be a preferred global supplier of energy because of its climate and environmental record, 69 percent said they have personally benefited from the oil and gas sector, and almost three of four agreed that Canada's oil and gas sector helps fund important social programs like health care and education. Seventy percent of respondents agreed that resource development could help alleviate systemic poverty within Indigenous communities, and 66 percent supported Canada's role as a global oil and gas supplier.

The same year, a study by Climate Access (a non-profit organization focused on "building political and public support for climate and clean energy solutions"), saw some evidence that Canadians want action on climate change as a here-and-now problem. But as Climate Access put it, "Those high levels of concern don't automatically translate into a social mandate for more aggressive action. . . . Most supporters fall into a muddled 'moveable middle.' Almost half of the Canadian public are worried but don't really understand climate change or what needs to be done about it. There's a core group (about 25% of the public) that are truly "Alarmed" and then a much larger group (about 45% of Canadians) that are "Concerned" but not very engaged. "This large moveable middle means that demand for action is much lower than concern. And it means public support isn't reliable when difficult, real world decisions come around." [72]

Of course, federal government policy is influenced by many forces, including public opinion, how elites feel, the pressure of large events (like weather catastrophes), and provincial governments. How will all these forces come together, and what will come out of them in terms of policies that impact the different sectors of the Canadian petroleum industry?

What part of all of all this *sturm und drang* is simply noise, and what part is real or impending change with an immediate impact on people, investment, and the future of the oil sands?

Are we at the beginning of the end of the fossil fuel era, or the end of the beginning?

First, let's start by establishing where we are today.

World Demand for Energy Is Increasing

There's a direct relationship between

improved living standards and the availability of cheap and reliable energy for electricity, transportation, heating, and industry. Access to efficient, cheap, and reliable energy is linked to improved living standards, including health, education, and food production and preservation. Additional demand for electricity in developed countries comes from computerization and automation.

Two and half billion people—a third of the human race—lack clean fuel for cooking and heating. The World Health Organization says 2–3 million people a year die from indoor air pollution, which is the consequence of burning wood, plant waste, and dung for fuel. Demand for all forms of energy, including fossil fuels, is driven by this powerful need.

If climate policy doesn't address this need, it'll be futile. Worse yet, it will discriminate against those struggling to better their lives.

Worldwide, the fossil fuel industry will be with us for a long time yet. Even the most reviled part of it—coal—is already enjoying a new lease on life as a result of two factors: India and China have enormous coal reserves and huge populations clamoring for more electricity, and pressure to close nuclear plants increases pressure to build new coal-fired electricity plants. The closing of nuclear plants in Japan and Germany led to increased demand for coal in both countries. By 2030

Japan—which had earlier pledged to cut coal consumption for 10 percent of its energy—now expects (because of the shutdown of nuclear plants) to produce 26 percent of its electricity from coal by 2030. Despite vague commitments to phase down coal production and use, China is building many more coal fired power plants (as well as increasing renewables, nuclear, and the importation of natural gas) and is the world's largest producer of coal-fired electricity.

Pressure to cut back oil and gas production in the Western world is being met with countervailing pressure to increase production in the developing world. Realistic climate change analysts are—at least privately—increasingly skeptical that net zero will be achieved by 2050.

- The world currently consumes about 90 million barrels of oil or oil equivalent a day. The world currently consumes about 90 million barrels of oil or oil equivalent a day. Most forecasts say demand for oil will increase slowly until the middle of the century at least.

- Coal currently accounts for one-third of world electrical generation.[73] Despite climate conference pressure to cut back on or eliminate coal production, demand for coal continues to increase. China accounts for 50 percent of world coal consumption and is continuing to build coal-fired power plants as well as import natural gas from Russia, the Middle East, and other foreign suppliers, including, potentially, Canada.

- Thanks largely to the stunning growth of the oil sands industry between 1990 and 2010, Canada is the fourth-largest producer of oil and natural gas liquids in the world.

10 Largest Oil Producers and Share of Total World Oil Production in 2020		
Country	Million barrels per day	Share of world total
United States	18.61	20%
Saudi Arabia	10.81	12%
Russia	10.50	11%
Canada	5.23	6%
China	4.86	5%
Iraq	4.16	4%
United Arab Emirates	3.78	4%
Brazil	3.77	4%
Iran	3.01	3%
Kuwait	2.75	3%
Total top 10	67.49	72%
World total	93.86	

Source. IEA World Energy Outlook, 2021
https://www.eia.gov/tools/faqs/faq.php?id=709&t=6

- Canada's oil (predominately from the oil sands) supplies Canadian markets west of Quebec; the rest—3.7 million barrels a day—is exported to the United States, making Canada the largest supplier of oil to that country.
- There are 17 refineries in Canada that have a collective crude oil–refining capacity of 2 million barrels per day.
- World demand for natural gas is expected to increase 22% by 2040, driven by rapidly

expanding Asian economies according to forecasts by the International Energy Agency. Canada's natural gas is uniquely positioned to meet that growing energy demand by developing a liquid natural gas industry. In recent times, US-fracked natural gas has pushed back Canadian suppliers. Overall, natural gas exports to the United States have dropped 22 percent in 10 years, a fact that's been the driving force behind building gas pipelines to Canada's West Coast.

- Despite having the world's third-largest oil reserves, Canada imports oil from foreign suppliers. Currently, more than half the oil used in Quebec and Atlantic Canada is imported from foreign sources including the United States, Saudi Arabia, Russia, the United Kingdom, Azerbaijan, Nigeria, and Ivory Coast. In 2019, Canada spent $18.9 billion to import foreign oil. Canada imported more than 660,000 barrels of oil in 2019.

10 Largest Oil Consumers and Share of Total World Oil Production in 2020		
Country	Million barrels per day	Share of world total
United States	20.54	20%
China	14.01	14%
India	4.92	5%
Japan	3.74	4%
Russia	3.70	4%
Saudi Arabia	3.18	3%
Brazil	3.14	3%
South Korea	2.60	3%
Canada	2.51	3%
Germany	2.35	2%
Total top 10	60.69	60%
World total	100.23	

Source. IEA World Energy Outlook, 2021
https://www.eia.gov/tools/faqs/faq.php?id=709&t=6

- While renewable energy is growing rapidly, it won't completely supplant oil and gas as an energy source in Canada for many years, certainly not in this century. Both wind and solar are low-energy-density fuels and their contribution to reliable, low-cost electrical energy will be limited. For one thing, our geography and climate pose major challenges to their widespread adoption. Fortunately, Canada has substantial hydroelectric, nuclear, and fossil fuel alternatives.

- Petroleum in all its sources—conventional oil, oil from the oil sands, natural gas, and natural gas liquids—is a pillar of the Canadian economy and supports an enormous number of direct and indirect jobs as well as government revenues that support our health care, education, and social infrastructure. The benefits are national and not limited to the western provinces.

Having said all that, when it comes to assessing the future of the fossil fuel industry, not all fossil fuels are the same, and the prospects for each depend on what part of the industry we're talking about. Canada's fossil fuel industry is made up of numerous sectors; each one faces different prospects.

The Oft-Forgotten and Most Vulnerable Fossil Fuel: Coal

The thermal coal industry was the first target of the climate change movement in both Europe and the United States. Coal is the highest greenhouse gas emitter of the industry and, let's face it, the one least liked by the public. As a power plant fuel it's easily replaced by natural gas, which is a much lower emitter. In Canada, thermal coal is on the way out: only 8.5 percent is used as power plant fuel.

It's worth noting that in the rest of the world, thermal coal is still a very large fuel source, and electrical generation and coal consumption is rising—not falling—in places like India and China. China relies on coal power for approximately 70–80 percent of its energy, with 40 percent used for the industrial sector and the remainder to generate electricity. By

2010, China comprised 48 percent of global coal consumption. India has similar ambitions.

It's an ugly fact, but there it is.

Canada's a significant producer of the so-called other form of coal—coking coal, sometimes called steelmaking coal. Most of it comes from British Columbia and Alberta and is exported to steel mills in Asia. Some 31 million tonnes a year of it, to be exact, with its own West Coast port at Roberts Bank, British Columbia. The industry is a major employer is southeastern British Columbia, and shutting down the industry would turn Fernie and Sparwood into ghost towns and create a significant political issue in the province. Besides, Canadian supply would be easily replaced by Australia's; the country has huge reserves, thus rendering nil any net reduction of emissions to the atmosphere.

At present, there's no practical replacement for coal in steel making. Hydrogen is being explored but is years away. And so there the steelmaking coal industry sits, glowing with menace.

Conventional Oil

Conventional oil production peaked in the early 1970s, and 90 percent of Canada's conventional oil reserves have been produced, so the country's conventional oil industry is a sunset industry. We just have to remember that it's a long sunset. We'll probably still be producing conventional oil for decades, although it will decline to a dribble.

Canadian oil production: conventional crude oil in red, and total petroleum liquids, including from oil sands, in black. (Figure source: Wikiwand, Petroleum industry in Canada - Wikiwand.)

From the point of view of climate policy, most of the conventional oil industry's emissions are already in the atmosphere, and little would be gained by shutting down the industry, which is already on the downslope anyway. Shutting it down would also impede the cleanup of its legacy oil wells, which dot Western Canada.

Natural Gas

Solar and wind power are starting to play a role in electricity generation but will not replace natural gas as a baseload fuel (they can't, given that both solar and wind are intermittent, not to mention expensive). And they're not expected to replace natural gas as of yet. Solar and wind will supplement natural gas and, of course, indirectly add their own greenhouse gases to the atmosphere (since they are manufactured from minerals that have to be mined and processed, which consumes energy). Base-load power, without which electricity networks break down, must be supplied by either fossil fuels or nuclear energy. Which is one reason why an all-renewables energy system is an arcadian myth, at least in this century.

That said, the prospects for Canadian natural gas are potentially good if pipelines can be built to get it to an increasingly natural gas–hungry world. Canada has a lot of natural gas and is a significant world player in the space. There's a large and growing export market for natural gas via pipelines and liquid natural gas carriers. Europe is a major natural gas consumer (mostly via pipelines from Russia), and Asia is as well. Reacting to Russia's invasion of Ukraine, European governments are taking steps to severely restrict natural gas imports from Russia.

Natural gas is the only fuel making it even close to possible for Europe to phase out coal and nuclear and embrace renewable energy. Without imported natural gas, Europe's economy would teeter, and this will be true for many, many years. These facts are not necessarily palatable to Europe's Greens, and that complicates getting to a rational energy policy there.

Canadian natural gas was not the first hydrocarbon to be targeted by climate change activists—US nuclear, coal, and fracked oil and gas shared that fate—and a reasonable person might think environmentalists would cut natural gas some slack in light of the fact that its emissions per unit of production are the lowest of any hydrocarbon and have made possible significantly reduced emissions in the electric utility industry. Despite that, natural gas has come in for more than its share of ideological opposition.

Within Canada, attacks on natural gas production have occurred chiefly in Quebec, which has banned development of its significant natural gas reserves in the St. Lawrence Valley, and British Columbia, where bitter environmentalist opposition has been mounted against the proposed Coastal GasLink pipeline to supply a liquid natural gas export terminal in Kitimat.

For a variety of good political reasons, the Government of British Columbia strongly supports Coastal GasLink, and the odds favour its eventual completion, even if a handful of Wet'suwet'en traditional chiefs and local green activists have to be dragged to the official opening kicking and screaming. As this was written, masked terrorists have attacked the Coastal GasLink pipeline workers. As with the Transmountain Pipeline, efforts are underway to increase Indigenous ownership of Coastal GasLink, and this will strengthen the odds of its completion, thus supporting significant new

Canadian natural gas exports to Asia.

In Quebec, natural gas developers, led by Questerre Ltd., are working to gain government approval for a circular economy approach to natural gas development that would be effectively net zero and enable Quebec to develop its reserves *and* export natural gas to Europe, thereby helping to reduce, but eliminate, Europe's dependency on fossil fuels. But the Quebec government succumbed to knee-jerk environmental opposition and rejected Questerre out of hand. The government's position, apparently, was that it was spinach, and so to Hell with it. At the time of this writing, the policy is unchanged.

Meanwhile, both the natural gas industry and the Alberta government are talking about transforming part or all of the natural gas industry into a world-class producer of the ultimate net-zero fuel: hydrogen.

How Natural Gas Can Power the Hydrogen Transition

In November 2021, Alberta Energy published a major policy document entitled *The Alberta Hydrogen Roadmap*, with a plan "to integrate hydrogen with the province's existing energy system and propel Alberta into the global hydrogen economy," thus making Alberta "a major player" in that economy.

Alberta is currently Canada's largest producer

of hydrogen, creating 2.4 million tonnes a year from natural gas feedstock. The *Roadmap* states that "Alberta's ambition is to reduce the carbon intensity for existing industrial hydrogen production as well as create demands markets for heating, transportation, power generation and supply, and export."

At the present time, the global consumption of hydrogen is about 90 million tonnes a year. Bloomberg and the International Energy Agency forecast that by 2050, this will rise to almost 700 million tonnes a year—24 percent of global energy demand—and generate nearly a trillion Canadian dollars in sales. Alberta—and perhaps ultimately British Columbia—aspires to become a major player in that market by producing clean hydrogen mostly from natural gas.

There are many challenges to overcome. The plan envisages Alberta exporting about 100 million tonnes of hydrogen by 2050 (14 percent of expected global consumption) but does not estimate what amount of capital investment will be required in Canada to achieve that goal. Nor does it suggest where that capital will come from or how all the many changes in technology, supply chains, and workforce training will be needed (not to mention new carbon capture and underground storage capacity).

Then, there's the question of whether a profitable industry can be built to produce and market hydrogen. Natural gas producers and distributors aren't a social service agency. Can a profitable business be built around hydrogen production at this scale? What's the business model?

Another obstacle is certain to be the ideological fixation of the climate change movement, which is already lining up to oppose the use of natural gas as a hydrogen feedstock. Parts of the movement grimly oppose "blue" hydrogen—hydrogen produced from natural gas—and are prepared to accept only "green" hydrogen, which is produced by the electrolysis of water using renewable electricity. The economics and technology of both are changing all the time.

Blue hydrogen, if it's coupled with carbon capture and storage, should be competitive in price with green hydrogen. Alberta's gambling that the global energy market may have room and need for both.[74]

At this point, the *Roadmap* is only aspirational. It's not a Government of Alberta plan backed up with budgets and business plans. It only hints at but does not address in any detailed way the many enormous gaps and challenges that stand in the way of achieving its goal. It offers some hope for people in and dependent on the natural gas sector, but so far, hope must be cautioned by two large and unanswered questions. Can the *Roadmap* be traversed by 2050, and at what cost?

The hydrogen highway, if it's built, would

appear to offer most of its succor to the natural gas industry, not the oil sands (natural gas is a much more plausible feedstock for hydrogen than bitumen). Some part of oil sands bitumen production could in theory be diverted to hydrogen production, but natural gas makes much more sense as a feedstock.

The Oil Sands Industry

And so we turn to the industry that is the subject of this book and by far the most important sector of Canada's fossil fuel complex: the oil sands. Here, the crystal ball becomes even cloudier.

By 2021, the oil sands companies—and the Alberta government that was and is their effective partner—were virtually encircled. The industry's attempt to boost exports to the US Gulf Coast, Asia, and even Eastern Canada had been defeated by the US foundation-funded climate change movement. Faced by international treaty requirements to drastically reduce greenhouse gas emissions, a federal government half-heartedly supporting the industry while also appeasing, if not embracing, its enemies and an Alberta government justifiably concerned about the health of the province's most important industry, not to mention the well-being of hundreds of thousands of people and communities dependent on it, the oil sands industry, still a very significant force in the world, sailed into uncharted waters.

In politics, war, and business, when you're facing a strong, hostile force, you have three choices: pretend you're not and hope the adversary will go away (in psychology, this is called "denial"), hunker down and fight to preserve the status quo, or adapt to the change. True, there's a fourth choice: throw yourself on your sword. But I'm assuming that no one who cares about the well-being of Canada would entertain that option.

So as things now stand (February 2022), the oil sands industry and its partner, the Alberta government, must choose between three strategic options. Let's give them a look.

1. Denial: Hope that the Climate Change Movement Will Get Tired and Go Away

This is the scenario of old King Canute trying to sweep back the tide with a broom. Or in Hollywood scriptwriter terms, it's the "wake up, darling, you were having a nightmare" scenario.

During the first decade of the century, denial might've characterized the thinking of some people in the energy industry as well as the Alberta political and business class. But is denial a basis for a realistic plan? Surely not.

You'd have to imagine a change in federal politics sufficient to return a Conservative government (led by a second coming of Stephen

Harper) to rewrite the energy script. You'd need to imagine the decarbonization movement running out of steam on the immensity of the challenge of unwinding the world economy's enormous dependency on fossil fuels.

You'd need to forget that powerful reform movements can take a very long time to wind down. The real Crusades lasted 108 years, and Europe's "Kingdom of Jerusalem" lasted almost 200. The Prohibition movement lasted more than 50 years, and parts of it survive in the form of the new Puritanism.

It's true that the world is still consuming 90 million barrels of oil a day and seems likely to consume even more in the future. The developing world is only starting to produce and consume the amount of energy it'll need to dramatically improve living standards, and renewables are unlikely to slake that thirst—and there are many reasons for that. Renewable energy has a *very* long way to go to make a significant dent in the world's energy needs. Since it's unreliable (when the sun doesn't shine and the wind doesn't blow) and since the developing world has a continuing and legitimate need for the energy that fossil fuels can provide, we *might* reach a point at which world leaders decide some or many of the forecasts of climate catastrophe have been overblown and unrealistic.[75] As this was written, in reaction to the war in Ukraine, there are signs that the realities of energy systems are causing European governments to put climate-change goals on the back burner.

Despite all of that we need to remember the Damon Runyan's comment about boxing: "A good big man won't always beat a good little man, but it's the way to bet."

For the previous two years, the COVID-19 pandemic has pushed climate change out of the headlines (or at least further down the front page). But that hasn't stopped international climate conferences from happening. It's possible other developments—new pandemics, another recession, refugee crises, wars on the periphery of Russia or China—will again push climate change out of its leading role in the media. Again, none of these scenarios seems likely enough to place your bets on.

Is it possible, in theory, that a Canadian political leader could decide to simply cut off Canadian oil and gas production for political reasons like satisfying Canada's international "climate commitments." Such a step would require choking off or slowly strangling Canada's most important export industry and energy provider in order to satisfy an evanescent "world opinion" and it would trigger a very angry popular reaction from western Canada. No major national leader has been willing to emasculate a domestic energy industry, to date, and as we have seen, developments from the war in

Ukraine have pushed European leaders in the opposite direction.

2. Hunker Down and Fight to Retain the Status Quo: Hold onto the Customers and Markets You Have, Squeeze the Most Out of Every Investment Dollar, and Do What You Can to Remain a Viable Business

What this option boils down to, would be accepting that the fossil fuel industry is destined to be a sunset industry—and to drag out the sunset for as long as possible. Continually squeeze down costs to protect margins, boost revenues where you can, and get the most out of a dying asset base.

Churchill once described Labour leader Hugh Gaitskell as "a desiccated calculating machine." This strategy might appeal to an oil sands CEO "channeling" Hugh Gaitskell but it would be a very hard swallow for the tens of thousands of people who work in the industry or supply it, the people who live in oil sands communities, and the millions of Canadians who benefit from the economic opportunities and tax revenues the industry provides.

And that's not to mention the political consequences. The last time someone named Trudeau brought in a National Energy Program that sent Alberta into a generation-long recession, it fanned Alberta separatism, led to the Reform Party, and overturned federal politics. If westerners were to conclude that their economy has been thrown to the wolves to satisfy some vague international goal—and many of them suspect that already—the political consequences wouldn't be pretty.

Which brings us to the third choice.

3. Adapt: Accept that We're Moving toward a Decarbonized World while Recognizing that It Will Probably Take the Rest of the Century to Get There. Commit to as Much Carbon Removal as Possible in the Near Term (by 2050) while Defending the Industry's Economic Viability in the Shorter Term. Start Reinventing the Industry by Maximizing Hydrogen Production and Developing Technology to Produce New Products from Carbon Beyond Fuels.

I have no inside information on the thought process of executives in the oil industry or Alberta politics, but based on public evidence—plus reasoning—I suspect this option is the one currently on the table.

If I'm right, it'll unfold in two directions:

1) The oil sands will take steps to dramatically reduce its greenhouse gas footprint. In the short term (2022 to the mid to late 2030s), it'll use carbon capture and underground storage. In the longer term (mid-2020s to 2050 and beyond), it'll do it by diverting a growing amount of bitumen away from fuel production

and into the production of new carbon products that aren't fuels.

2) The natural gas industry—acting in parallel with the oil sands industry—will migrate a larger and larger portion of its production to the manufacture and distribution of hydrogen.

Step One: Reducing the Oil Sands' Greenhouse Gas Footprint

In June 2021, the five major oil sands companies—Suncor, Syncrude, Canadian Natural Resources, Cenovus Energy, Imperial Oil, and MEG Energy, who represent 80 percent of Alberta's production—joined with the Alberta government to announce the creation of *The Oil Sands Pathways to Net Zero Initiative.*

Pathways to Net Zero is both an agreement within the oil sands industry and a commitment to government (both the Alberta government and, tacitly, Ottawa) to embrace two possible solutions to a net-zero carbon future. Its principal thrust is a commitment to reduce the upstream emissions of the oil sands industry to zero by 2050 "to help Canada meet its climate goals, including its Paris Agreement commitments and 2050 net zero aspirations." ("Upstream" emissions are GHGs produced by the extraction of bitumen from the oil sands. They do not include "downstream" emissions produced by the burning of bitumen products – like gasoline – by cars, trains, industry and households).

- "A core infrastructure corridor linking oil sands facilities in the Fort McMurray and Cold Lake regions to a carbon sequestration hub near Cold Lake via a CO2 trunkline. The trunkline would also be available to other industries in the region interested in capturing and sequestering CO2. There's also potential to link the infrastructure corridor to the Edmonton region."[76]
- "Deploying existing and emerging GHG reduction technologies at oil sands operations along the corridor, including [carbon capture and underground storage] technology, clean hydrogen, process improvements, energy efficiency, fuel switching and electrification.
- Evaluating, piloting and accelerating application of potential emerging emissions-reducing technologies including direct air capture, next-generation recovery technologies and small modular nuclear reactors."[77]

Putting such a system in place will require an large amount of capital—the figure of $30 billion has been mentioned—and a breathtaking level of co-operation among companies and between companies and government. The announcement alluded to "welcome announcements" from the

Government of Canada and the Government of Alberta on important support programs for emissions-reduction projects and infrastructure. Collaboration between industry and government will be critical to progressing *The Oil Sands Pathways to Net Zero* vision and achieving Canada's climate goals.

It added, "The Pathways initiative is ambitious and will require significant investment on the part of both industry and government to advance the research and development of new and emerging technologies" and that it'll "engage with local indigenous communities in northern Alberta to make this ambitious, major emissions-reduction vision a reality so those communities can continue to benefit from Canadian resource development."

Will this initiative be a game-changer for Canada's climate change policy? Two big questions need answering.

The first is what part of the cost—the $30 billion—will be borne by government, including the federal government. The launch announcement said industry would also be looking for government "to develop enabling policies, fiscal programs and regulations to provide certainty." These would include dependable access to carbon sequestration rights, emissions reduction credits, and ongoing investment tax credits. In practical terms, rumours suggest that the industry will be looking for an 80 percent contribution from government, which is a large swallow, especially in light of the fact that the climate change movement will oppose it with all its might as a sell-out to the fossil fuel industry.

The second question will be the climate movement's stubborn, largely ideological opposition to carbon capture and underground storage system based largely on the belief that it's a stall tactic the industry is using to stave off shutdown. But the technology, if carried out, could potentially reduce the oil sands' upstream production emissions of CO2 by about 30 percent (process changes could conceivably raise this figure higher).

Will federal politicians (Liberals, basically) accept a 30 percent reduction in greenhouse gas emissions as a sufficient industry contribution to federal climate goals...or will they view it as a down payment toward the shutdown of the industry?

Step Two: Divert a Growing Part of Oil Sands Bitumen into a Feedstock for Creating Carbon Products that Aren't Fuels

The second part of a plan to drastically reduce oil sands emissions footprint takes us into a world of imagination that almost looks like science fiction: it is diverting a growing proportion of bitumen production to the production of new carbon-based products like nanotubes, graphene, and other advanced materials and products that can make transportation more energy efficient, infrastructure less energy intensive and more durable, and renewable generation and energy storage more economic.

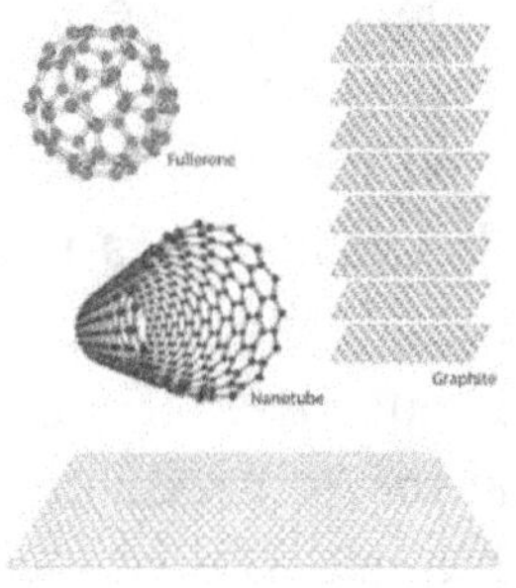

Graphene. (Figure source: https://www.sigmaaldrich.com /CA/en/technical-documents/technical-article/materials-science-and-engineering/bioelectronics/graphene-in-biotechnology.)

In November 2021, Alberta Innovates, a provincial agency that embraced the old Alberta Research Council, AOSTRA, and other provincial agencies, published *Bitumen Beyond Combustion: How oil sands can help the world reach net-zero emissions and create economic opportunities for Alberta and Canada.*[78]

This report proposed that Alberta pivot over time from producing oil sands fuels and petrochemicals to producing "a value chain of advanced materials and products to make transportation more energy efficient, infrastructure less energy intensive and more durable, and renewable generation and energy storage more economic." Combined with carbon capture and storage, this would turn the oil sands industry into a fully net-zero "multi-billion-dollar clean tech engine that will propel a net-zero economy" in which "the carbon from the bitumen remains sequestered within the products and not released. In fact, the carbon becomes an asset."

The environmentally safe, no-emission carbon products referred to include carbon fibre, asphalt binder, activated carbon, graphene, carbon nanotubes, metal carbides, and synthetic graphenes.

By diverting a portion of bitumen's heavier fractions from fuel to environmentally safe carbon-based products, the paper argued, Alberta would both reduce downstream emissions and create a new value stream for the economy. "Regardless of scenario, the production of carbon fibres, asphalt and high-value carbon materials can contribute $3 billion of revenue in 2030 by converting only one percent of bitumen production."

Bitumen Beyond Combustion diplomatically mentioned the elephant in the room. Namely, that the Government of Canada's strengthened climate plan, *A Healthy Environment and a Healthy Economy*, talked about everything under the sun -- energy efficiency, hydrogen and biofuels production, battery materials development, electrification in transportation, and carbon capture and storage underground – but was deafeningly silent on " a vision for the oil sands industry in a net-zero world,"

Bitumen Beyond Combustion asked the "elephant" question: How can the oil sands help the world reach net-zero emissions and create economic opportunities for Alberta and Canada?"

Bitumen Beyond Combustion wasn't some exercise in blue-sky visioning. It incorporated serious science, stakeholder consultation, and big-picture economic analysis into a futures document that deserves respect. However, at the time of this writing, it has yet to be adopted by either Alberta or the federal government. It remains a creative aspiration, one future among many. Whether it will become the basis for a whole new industry supporting the Albertan and Canadian economies is a question.

Another question is whether the climate change movement and its supporters in the political system will agree that zero carbon emissions from the upstream oil sands (that is, from the production of bitumen and synthetic crude) will be enough. The most zealous climate change activists—and they are very clear about this in their writings and pronouncements—are absolutists who'll accept no compromise to the goal of total replacement of fossil fuels by zero-carbon energy sources.

In early 2019, some 600 environmental

groups submitted a letter to the US House of Representatives that stated that the United States must shift to "100 percent renewable energy power generation by 2035 or earlier." The same letter said that "any definition of renewable energy must . . . exclude all combustion-based power generation, nuclear, biomass energy, large -scale hydro and waste-to-energy technologies," and the new electric grid must have the "ability to incorporate battery storage and distributed energy systems that are democratically governed."

Similar statements pepper the policy pronouncements of Canada's centrist, leftist, and green parties and constitute a kind of energy policy mantra. They provide moral support for radical actions to derail fossil fuels. In January, 2022, David Suzuki, one of the early leaders of the environmental movement, mused about what may happen if the movement doesn't get its way. "We're in deep, deep doo-doo," he said an Extinction Rebellion protest on Vancouver Island. "This is what we've come to. The next stage after this, there are going to be pipelines blown up if our leaders don't pay attention to what's going on."[79] (Wink-wink, nudge-nudge). Only weeks following this, balaclava-wearing "protesters," wielding axes, attacked a work camp of Coastal GasLink pipeline workers who had to summon police for protection.

The industry has taken a statesmanlike

stance, stating that "the most effective way to address climate change is by developing and advancing new technologies and that this unprecedented challenge can and will be solved by Canadian ingenuity, leadership and collaboration."

* * *

And so, at last, we return to the future of the Canadian oil industry. It's a future that is still being shaped by oil industry thinkers, politicians, scientists, economic forecasters, policy wonks, and climate change activists in Europe and North America.

The future of the industry is like a highway being built through a forest. Will industry and government push the asphalt farther into the trees, and if so, where will it lead? A sunny upland . . . or a precipice?

Where will the visions of Sidney Ells, Karl Clark, Bob Fitzsimmons, Ernest Manning, J. Howard Pew, Frank Spragins, Peter Lougheed, and so many others over a century—not to mention the labours of tens of thousands of ordinary Canadians—finally lead?

Will Canadian industry and government channel the vision of those pioneers and invest in the enormous innovation that will be required to give the oil sands a new beginning?

It's an industry that has overcome many doubters and skeptics in the past. Maybe

creativity and innovation are in its genes.
Once a great notion . . .

EPILOGUE

We enter the New WestJet Aerospace terminal, board the *Spacetime Venturer*, and fasten our seatbelts. The captain comes on the viewscreen introducing himself.

"Lou L'Hirondelle," he says, smiling. "I understand you want to go to the end of the 21st century."

"We want to see the future of the oil sands I say. "Like you took us back to birth of it, but the other direction."

"That's the plan," he says. "Government regulations say I need to remind everyone that this is a simulation made possible by artificial intelligence. In other words, it's based on probabilities and forecasts. It's perfectly safe and won't damage the environment." He smiles broadly. "Please keep your hands on the armrests and resist the temptation to reach out the 'windows."

We laugh nervously.

"If you agree, please click the green button on your seat rest, and then sit back and enjoy the experience."

I think to myself, *Make it so, Number One.*

"Okay," the captain says. "Hold onto your hats."

Outside the window—I remind myself it's just a screen with a projected image—the view begins to spin past. The sun flickers like a faint strobe light. The landscape dissolves into a blur, from green to brown to white and back again, 1,000 times a second. I feel faint and close my eyes tightly. After a time—it's hard to say how long— it feels like we're slowing.

"Thought you might like to detour past the big fusion plant near Cold Lake," the captain says over the speaker. .

We look down and see an industrial complex next to a huge concrete dome near the shore of an immense stretch of blue. Power lines and towers radiate away to the west and south.

"They call it UniPower," the captain says. "It supplies baseload electricity for when the solar and wind farms aren't working to all of Alberta and Saskatchewan. They built it here so they can extract deuterium from the lake and use the lake itself to cool the reactor. They say there's enough deuterium in the lake to run it for 1,000 years. Anyway, you said you wanted to look at the oil sands. I'm dialing in the summer of 2199. Hold on."

It feels like we're banking again, and the world spins by in another blur. Then, finally, we slow. Beneath us is the familiar dark green of the vast northeastern boreal forest. It looks pretty much as I remember, but here and there are patches of lighter, more vibrant green. I ask about that.

"That's one of the effects of climate change," the pilot says. "Warmer soil and more CO2 in the atmosphere means food for the poplars and grasses. "Course, there are more forest fires now too."

Beneath us, the wide, brown Athabasca River winds its way north, broad and slow. We pass over the old divided highway running toward Edmonton; there are fewer trucks on it than I remember, electric ones if I can judge by the absence of exhaust stacks.

We turn and fly north along the river, then slow and descend to within a few metres of the riverbank. "That looks like bitumen dripping down the riverbank," I say. "Am I seeing things?"

"No, you got it right," the captain says. "Course, they don't mine much bitumen anymore, but there's still lots of it in the ground. You said you wanted to see Syncrude, or at least, where Syncrude used to be. Hold on."

We speed north and hover over a contoured grassy field; a herd of bison scatter as we pass over them. "That's Don Thompson Park," the captain says. "Over there to the right, that's Lake Spragins."

We skim over a wide pond fringed with bullrushes and willows. The water is blue with just a hint of a purple sheen.

"The AI says it's the original tailings pond, reclaimed. Apparently, a little residual oil floats up from time to time. The companies are still working to get all the clays to settle on the bottom. There's some aquatic life—bullfrogs and stuff—but apparently, it'll be a while yet before you can fish in it."

I touch my mike button. "What about the Syncrude plant?" I ask. "Anything left of it?"

"Let's take a look," he says.

We lift and speed north. Suddenly, we're over the ruins of buildings and old roads.

"That's what's left of it," the captain says. "They took down all the piping and steel girders and stuff and sent it to recyclers. About 30 years ago. They're still producing oil in situ south of McMurray. Cenovus uses it for petrochemicals and carbon fibre and stuff. I hear some of it's being used on the Mars ships."

We hear the captain talking to someone on the radio.

"Sorry, folks, the system tells me we've only got another 15 minutes of our AI quota, so we've got to head home. Please secure your chairbacks and tables in the upright position. And thanks for flying with New WestJet Aerospace."

---

[1] See Brian Fagan (ed.), *The Complete Ice Age: How Climate Change Shaped the World* (London: Thames and Hudson, 2009).

[2] Ross Anderson, "America's Atlantis," *The Atlantic*, October 2021; Tom D. Dillehay, "Peopling of the New World," *Encyclopedia of Archaeology*, 2008.

[3] For a detailed account of this period, roughly from the beginning of the 20th century until the late 1950s, see Peter McKenzie-Brown's *Bitumen: The People, Performance and Passions behind Alberta's Oil Sands.*

[4] The hippo said, "No way. We'll get part way across, and you'll sting me." To which the scorpion replied: "Why would I do that? If you drown, I drown." That made sense to the hippo, so the scorpion climbed on, and they started across. In the middle of the river, the scorpion stung the hippo. "Why?" the hippo gasped as they both sank. "*C'est le Congo*," said the scorpion.

[5] See Scott Allen, "The Greening of a Movement," *Boston Globe*, October 19, 1997; Douglas Jehl, "Charity Is the New Force in Environmental Fight," *New York Times*, June 28, 2001; Brian Seasholes, "The Green Pipeline," *Green Watch*, October, 2012.

[6] Personal communication with me in January 2021.

[7] A later president of Sun Oil said wryly,

"Bechtel didn't design it for Canadian conditions. We have this little thing called winter."

[8] See Oil Sands Oral History Archive at http://petroleumhistory.ca/oralhistory/.

[9] See http://www.history.alberta.ca/EnergyHeritage/sands/mega-projects/experimentation-and-commercial-development/the-winnipeg-agreement.aspx. For a highly critical interpretation of the Winnipeg Agreement and the way in which the public sector supported oil sands development. See also the Parkland Institute's report at https://www.parklandinstitute.ca/fifty_years_of_albertas_oil_sands. For a more positive interpretation see Ted Morton et al. at https://www.policyschool.ca/wp-content/uploads/2016/03/siren-song-economic-diversification-morton-mcdonald.pdf.

[10] Full disclosure: from 1972 to 1986, I was Syncrude's director of public affairs and responsible for the company's internal and external communications.

[11] Sally Denton, *The Profiteers: Bechtel and the Men Who Built the World*, (2017; repr., Toronto: Simon & Schuster, 2016).

[12] See Oil Sands Oral History Archive at http://petroleumhistory.ca/oralhistory/.

[13] Some writers contend they might've been joined by people from the Aleutians and even Polynesia, but this

hypothesis isn't widely supported by scholars. See Jennifer Raff, "Genomes Reveal Humanity's Journey into the Americas," *Scientific American*, May 2021, https://www.scientificamerican.com/article/genomes-reveal-humanitys-journey-into-the-americas/.

[14] For a detailed description of those days, see John J. Barr, *Dynasty: The Rise and Fall of Social Credit* (Toronto: McClelland and Stewart, 1975).

[15] Full disclosure: From 1972 to 1986, I was Syncrude's director of public affairs and took part in many of the events described in this book. Garvin was an important member of our team from 1973 until 1982.

[16] *Bush Land People* (Calgary: The Arctic Institute of North America, 1992) was Garvin's unique, respectful, and deeply personal portrait of many Indigenous families that lived in the northeast during the 1950s, 1960s, and 1970s. It provided and still provides a window into their lives and values. It was published a year after Garvin finished his work with Syncrude.

[17] See *Oil Sands Oral History Archive* at http://petroleumhistory.ca/oralhistory/.

[18] Sadly, Spragins was in the third year of a struggle against metastasizing cancer. He died two months after the official start-up of Syncrude in November 1978.

[19] Scott died in early 2021 at age 94. He and his wife Lilian were married for 73 years and had three sons, all of them accomplished. He

was a man of wide interests and broad intellect. He painted, travelled widely, skippered a good-sized sailboat until he was in his seventies, and never stopped learning. Even in the year before his death, he was still taking online courses in economics and management.

[20] See Harold G. Moore and Mike Guardia, *Winning When You're Outgunned and Outmanned* (CreateSpace Independent Publishing Platform, 2017).

[21] "The Challenge of the 1980s for Executives and Employees," Personnel Association of Toronto, April 21, 1978.

[22] The next two were John Elson, who built Northward Development Ltd., Syncrude's housing-development subsidiary, and Jonno Hanafin, who became manager of organizational development.

[23] See Chris Turner, *The Patch: The People, Pipelines, and Politics of the Oil Sands* (Toronto: Simon and Shuster, 2017).

[24] In 2001, the Organization Development Network held its annual conference in Vancouver and invited Brent Scott to speak. During the Q&A session, a member of the audience asked him how Shepard managed to persuade him to hire a 25-year-old with no experience for such a formidable job. Without missing a beat or cracking a smile, Scott replied "He was only 25?!"

[25] Drew Anderson and Allison Dempster, "Lougheed, Trudeau and the Notorious NEP," CBC.ca, https://newsinteractives.cbc.ca/longform/notorious-nep.

[26] Over time, Syncrude added two more mines, making its footprint the largest in the oil sands mining sector. In his time, Goforth only had to address the original mine.

[27] For a sense of the consequences of a tailings dam failure, see the picture of the Mt. Polley copper mine disaster on p. 119.

[28] This isn't intended as a criticism, as neither, for many years, did anyone else.

[29] The cokers weren't universally loved by Syncrude's process engineers because they were notoriously tricky to operate. One Exxon coker expert told his Syncrude counterpart, "There's a reason so few of these fuckers were ever built."

[30] For background on lichen monitoring see https://www.researchgate.net/publication/237154428_Lichens_and_sulphur_dioxide_air_pollution_Field_studies.

[31] This didn't prevent the public's misunderstanding of the issue. In the summer of 1984, Fort McMurray media were bombarded with calls from citizens who found a fine yellow dust coating their cars and sidewalks. "Is it acid rain?" several callers asked. It turned out to be spruce pollen from the surrounding forests.

[32] "The current ambient air quality monitoring data for the region show minimal

impacts from oil sands development on regional air quality except for noxious odour emission problems over the past two years. Control of NOx emissions and regional acidification potential remain valid concerns."

[33] Syncrude has introduced process improvements that have significantly reduced the amount of bitumen escaping the tailings pond. It's now thought that a high percentage of bitumen in the oil sands ore is now being recovered.

[34] Pat Leonard, "More Than 4 Billion Birds Stream Overhead During Fall Migration," *Cornell Chronicle,* https://news.cornell.edu/stories/2018/09/more-4-billion-birds-stream-overhead-during-fall-migration.

[35] It might've happened at some of the eight other tailings ponds in the region. Two years later, after an ice storm forced ducks onto Syncrude's Aurora mine tailings pond, the company had to euthanize 230 that had become oiled, managed to fly away, and were soon exhausted. Unable to fly any longer, they were found on a road.

[36] See Turner, *The Patch*, p. 208.

[37] I left Syncrude in 1986 and could only witness the incident as the rest of the world did, through the news media. As a crisis-management consultant for 25 years, I sympathize with both Syncrude and the premier and understand their deer-in-

headlights response. Having said that, both the company and the Alberta government were disappointingly caught off guard by an incident that they should've been prepared for.

[38] Bridget Mintz Testa, "Reclaiming Alberta's Oil Sands Mines," *Earth Magazine,* January 5, 2012, https://www.earthmagazine.org/article/reclaiming-albertas-oil-sands-mines.

[39] See Alberta Environment and Parks, "Total Area of the Oil Sands Tailings Ponds over Time," March 4, 2015, http://osip.alberta.ca/library/Dataset/Details/542.

[40] Canada Action, "How Much of the Oilsands Has Been Reclaimed?" August 20, 2019, https://www.canadaaction.ca/how_much_of_the_oilsands_has_been_reclaimed.

[41] For the full Royal Society of Canada expert panel report, see https://rsc-src.ca/en/environmental-and-health-impacts-canadas-oil-sands-industry.

[42] The year it promised and at the cost it promised, give or take a few hundred million dollars.

[43] Full disclosure: after the Owners announced that Syncrude's Edmonton head office would be closed, Shepherd asked me to relocate to Fort McMurray. For personal reasons, I declined and resigned with regret.)

[44] See Newell's interviews in the Oil Sands

Oral History Archive at https:// glenbow.ucalgary.ca/wp-content/ uploads/2019/06/Newell_Eric.pdf.

[45] See Oil Sands Oral History Archive at http:// petroleumhistory.ca/oralhistory/.

[46] See Pat Lee Shipman, "The Bright Side of the Black Death," *American Scientist*, accessed March 22, 2022, https://www.americanscientist.org/ article/the-bright-side-of-the-black-death.   As my grandmother used to say, "It's an ill wind that doesn't blow somebody good."

[47] Matthew Nisbet, a former journalism school dean, quotes *The Guardian's* transformation in its own words. See "Sciences, Publics, Politics: The Trouble with Climate Emergency Journalism," *Issues in Science and Technology*, Vol. XXXV, No. 4 (Summer 2019). See also Andrew Sullivan, "I Used to Be a Human Being," in *New York Magazine*, September 2016.

[48] See Struan Stevenson and Tony Singh, *The Course of History: Ten Meals that Changed the World* (New York: Arcade Publishing, 2019).

[49] For a report of one improper oil industry cases, see "Bribery in the Oil and Gas Industry: the Panalpina Scandal," at https://www.whistleblowers.org/bribery-in-the-oil-and-gas-industry./_For a dramatized but not unbalanced portrayal of oil industry scheming in the Middle East in the 1990s, see the 2005 film *Syriana*, directed by Stephan Gaghan.

[50] Full disclosure: from 1972 to 1986, I was responsible for Syncrude public affairs, so I may be biased. See https://www.johnjbarr.com/.

[51] For a comprehensive history of the issue, see Alice Bell, *Our Biggest Experiment: An Epic History of the Climate Crisis* (Berkeley, California: Counterpoint, 2020).

[52] World Meteorological Organization, "Declaration of the World Climate Conference," at https://dgvn.de/fileadmin/user_upload/DOKUMENTE/WCC-3/Declaration_WCC1.pdf.

[53] See https://en.wikipedia.org/wiki/ExxonMobil_climate_change_controversy.

[54] United States Congress, "Sagan Tells Congress About the End of the World (1985)," *The Film Archives*, September 10, 2018, YouTube video, 2:45:25, https://www.youtube.com/watch?v=2xBequjDEX0.

[55] See Bjorn Lomborg, *Cool It: The Skeptical Environmentalist's Guide to Global Warming* (Knopf, 2013) and *False Alarm: How Climate Change Panic Costs Us Trillions, Hurts the Poor, and Fails to Fix the Planet* (Basic Books, 2020); Steven A. Koonin, *Unsettled: What Climate Science Tells Us, What It Doesn't, and Why It Matters* (Dallas: Ben Bella Books, Inc., 2021); Vaclav Smil, *Energy Transitions: Global and National Perspectives* (Santa Barbara: Praeger, 2017). For a summary of Smil's latest book, see Dale Eisler, "Is Vaclav Smil the voice of reason we all need to hear?" *Public*

Policy Forum, https://ppforum.ca.

[56] Dale Eisler, "Is Vaclav Smil the voice of reason we all need to hear?" *Public Policy Forum*, https://ppforum.ca._Vaclav Smil, *How the World Really Works: The Science Behind How We Got Here and Where We're Going* (Penguin Books, 2022).

[57] For a view of the players in the climate change movement, see "Who's Who at Cop26," www.theguardian.com/environment/2021/oct/11/whos-who-at-cop26-the-leaders-who-hold-the-worlds-climate-in-their-hands.

[58] Colin Kinneburgh, "The Fracktivists," *Dissent*, Summer 2015, https://www.dissentmagazine.org/article/anti-fracking-new-york-global-algeria-poland.

[59] Koonin, *Unsettled*, p. 97.

[60] United Nations Climate Change, "The Glasgow Climate Pact–Key Outcomes from COP26," The Glasgow Climate Pact – Key Outcomes from COP26 | UNFCCC.

[61] Ehsan Masood and Jeff Tollefson, "`COP26 Hasn't Solved the Problem'; Scientists React to UN Climate Deal," *Nature*, November 14, 2021, https://www.nature.com/articles/d41586-021-03431-4.

[62] In 2021, a Dutch environmental group sued Royal Dutch Shell to reduce its emissions beyond what the company had promised. The Dutch court agreed, obliging Shell to reduce its emissions by 45 percent at the end of 2030

compared to 2019. Whether this judgement will survive appeal is unclear. It's doubtful that courts in North America could be persuaded to follow the Dutch ruling, but you never know.

[63] See Royal College of Psychiatrists, "The Climate Crisis Is Taking a Toll on the Mental Health of Children and Young People, November 20, 2020, https://www.rcpsych.ac.uk/news-and-features/latest-news/detail/2020/11/20/the-climate-crisis-is-taking-a-toll-on-the-mental-health-of-children-and-young-people.

[64] For an account of their campaign, see https://www.nytimes.com/2021/06/09/business/exxon-mobil-engine-no1-activist.html.

[65] Terence Corcoran, "Inquiry Into Foreign Funding of Anti-Alberta Energy Campaigns Could Shake Up Enviro Charities," *Financial Post*, September 18, 2019, https://financialpost.com/opinion/terence-corcoran-inquiry-into-foreign-funding-of-anti-alberta-energy-campaigns-could-shake-up-enviro-charities.

[66] Stephan J. Allan, *Report of the Public Inquiry Anti-Alberta Energy Campaigns*, https://www.alberta.ca/public-inquiry-into-anti-alberta-energy-campaigns.aspx.

[67] Chris Turner, *The Patch: The People, Pipelines and Politics of the Oil Sands* (New York: Simon and Shuster, 2017), "The Climate Bomb," which is the whole idea that the climate is going to Hell and

everything we do in life will have to change to combat it.

[68] As stated in an email to me in April 2021.

[69] As opposed to a big lie. By the most objective measurement, that of the US State Department, depending on whether you count the greenhouse gas produced by burning the oil or emitted at every stage of a barrel of oil's life, from its production to its combustion, bitumen contributed 17 percent more greenhouse gas than oil from the Middle East. The problem with this argument was that the greenhouse gas content of oil sands crude has been steadily dropping, and even if refineries on the Gulf Coast stopped importing it, they'd simply replace it with oil from elsewhere and with similar greenhouse gas content. So the benefit to the Earth's atmosphere, in terms of emissions, would be essentially nil.

[70] Daniel Yergin, *The New Map: Energy, Climate and The Clash of Nations* (New York: Penguin Books, 2021).

[71] Chris Hatch, "What Do Canadians Really Think About Climate Change?" *Climate Access*, March 10, 2021, https://climateaccess.org/blog/what-do-canadians-really-think-about-climate-change.

[72] Hatch, "What Do Canadians Really Think About Climate Change?"

[73] BP Statistical Review of World Energy.

[74] The world is still struggling to produce a

taxonomy of environmentally acceptable forms of hydrogen. Hydrogen doesn't exist in pure form on Earth; it has to be liberated from hydrocarbons or produced by the electrolysis of water. Environmentalists refer to blue hydrogen (produced from natural gas) and green hydrogen (produced by electrolysis of water using renewable electricity). Since the production of blue hydrogen emits CO2, only green hydrogen (which emits no carbon) is thought to be permissible. In the future, we may hear about other forms of politically incorrect hydrogen (grey hydrogen? brown?).

[75] See Michael Shellenberger, *Apocalypse Never: Why Environmental Alarmism Hurst Us All* (New York: Harper Books, 2020).

[76] Emphasis added. It describes the proposed carbon capture and underground storage system "as similar to the multi-billion-dollar Longship/Northern Lights project in Norway as well as other [carbon capture and underground storage system] projects in the Netherlands, U.K. and U.S," see "Canada's Largest Oil Sands Producers Announce Unprecedented Alliance to Achieve Net Zero Greenhouse Gas Emissions," June 09, 2021, https://news.imperialoil.ca/news-releases/news-releases/2021/Canadas-largest-oil-sands-producers-announce-unprecedented-alliance-to-achieve-net-zero-greenhouse-gas-emissions/default.aspx.

[77] See "Canada's Largest Oil Sands Producers

Announce Unprecedented Alliance to Achieve Net Zero Greenhouse Gas Emissions."

[78] Alberta Innovates describes itself as "a provincial corporation delivering seed funding, business advice, applied research and technical services, and avenues for partnership and collaboration."

[79] Tyler Dawson, "'Pipelines Will Be Blown Up,' Says David Suzuki, If Leaders Don't Act on Climate Change," *Financial Post*, November 22, 2021, https://nationalpost.com/news/canada/pipelines-will-be-blown-up-says-david-suzuki-if-leaders-dont-act-on-climate-change.

REFERENCES

Barr, John J. *The Dynasty: The Rise and Fall of Social Credit in Alberta*. Toronto: McClelland and Stewart, 1975.

———. "The Time Of The Tar Sands," *Provincial Archives of Alberta*, December 9, 2019. YouTube video, 29.57. https://www.youtube.com/watch?v=xjFmOAKFLM0.

Denton, Sally. *The Profiteers: Bechtel and the Men Who Built the World*. Toronto: Simon & Schuster, 2017. First published Simon & Schuster, 2016.

Ells, S.C. *Preliminary Report on the Bituminous Sands of Northern Alberta*. Ottawa: Canada Mines Branch, 1914, https://archive.org/details/preliminaryrepor00cana/page/n5/mode/2up.

Fagan, Brian, ed. *The Complete Ice Age: How Climate Change Shaped the World*. London: Thames and Hudson, 2009.

Ferguson, Niall. *Doom: The Politics of Catastrophe.* New York: Penguin Press, 2021.

Fitzsimmons, R. C. *The Truth About Alberta Tar Sands: Why Were They Kept Out of Production?* Edmonton: R.C. Fitzsimmons, 1953.

Hunt, Joyce E. *Local Push – Global Pull: The Untold History of the Athabasca Oil Sands 1900-1930.* Calgary: Push-Pull Ltd., 2011.

Kinser, Stephen. "BP and Iran: The Forgotten History." *CBS News.* June 30, 2010.

Koonin, Steven A. *Unsettled: What Climate Science Tells Us, What It Doesn't, and Why It Matters.* Dallas: Ben Bella Books, Inc., 2021.

Lomborg, Bjorn. *Cool It: The Skeptical Environmentalist's Guide to Global Warming.* New York: Knopf, 2013.

———. *False Alarm: How Climate Change Panic Costs Us Trillions, Hurts the Poor, and Fails to Fix the Plan*et. New York: Basic Books, 2020.

McKenzie-Brown, Peter. *Bitumen: The People, Performance and Passions Behind Alberta's Oil Sands.* CreateSpace Independent Publishing Platform, 2017.

Harold G. Moore and Mike Guardia, *Winning When You're Outgunned and Outmanned.*

CreateSpace Independent Publishing Platform, 2017.

Peebles, Torren. "Development of Hubbert's Peak Oil Theory and Analysis of its Continued Validity for US Oil Production," 5 May 2017, https://earth.yale.edu/sites/default/files/files/Peebles_Senior_Essay.pdf.

Rasporich, Anthony W., ed. *The Making of The Modern West: Western Canada Since 1945.* Calgary: University of Calgary Press, 1984.

Roberts, Shane. *Separating the Sands: Karl Clark and Early Oil Sands Research in Alberta.* Thesis, University of Western, 2018.

Smil, Vaclav. *Energy Transitions: Global and National Perspectives.* Santa Barbara: Praeger, 2017.

————. *Power Density: A Key to Understanding Energy Sources and Uses.* Cambridge: The MIT Press, 2015.

————. *How the World Really Works: The Science Behind How We Got Here and Where We're Going.* Canada: Viking, 2022.

Turner, Chris. *The Patch: The People, Pipelines, and Politics of the Oil Sands.* Toronto: Simon and Shuster, 2017.

Yager, David. *From Miracle to Menace: Alberta, A*

Carbon Story. Victoria: FriesenPress, 2019.

Yergin, Daniel. *The New Map: Energy, Climate and The Clash of Nations*. New York: Penguin Books, 2021.

ACKNOWLEDGEME NT

This is not a definitive history of the oil sands, the people who made the oil sands useful to humanity, or the people who worked to prevent it. It is an attempt to provide a balanced perspective on all three, at a time when we are all being battered with too much rhetoric about these matters.

It was important to me to illuminate how the history of the oil sands has been driven by people, and how the development of the industry has changed the lives of people: scientists, business executives, indigenous people, working people of all kinds. If you are looking for pure heroes or villains, you've come to the wrong place.

For fourteen years, as Syncrude's first communication head I was privileged to know and work with some of the principal actors in the drama and participate in the great pivot point in the development of the Alberta oil sands. Now, late in my life, with a host of still-vivid memories and impressions, I am able to give an account of it.

The design and operationalization of the Syncrude complex deserves an honored place in the lexicon of Canadian achievements but has not been well told before (and today is largely taken for granted). The creation of Syncrude's human organization—its people, its culture, its way of dealing with employees and their families, Indigenous people and Canadians in general—by and large has not been reported or recognized by society. Perhaps not even sufficiently by the participants themselves (those few of us who are still around.)

I hope to provide some insight into what is today a highly controversial question, namely has the oil sands changed our society and the planet itself? How did the creation of the first great oil sands mega-project, Syncrude—both the human organization itself and the huge physical complex that the people of Syncrude built and operated—become a milestone in Canadian history? This book does not purport to be the final answer to these questions, which are vastly complex. I hope that you will think of it as an inside account of a business saga that was deeply important, not just to those of us who shaped it, but to our country and the world.

Many people provided me information and counsel during the extended writing of this book and want to acknowledge them. Preston

Manning was a valuable source on the creation of Great Canadian Oil Sands and the role played by his father , Ernest C. Manning, and J. Howard Pew. C.R. (Chuck) Collyer provided valuable perspective on the "big, tough, expensive job" of building Syncrude. Alex Gordon, Jonno Hanafin and Norman Cottee helped me tell important chapters in Syncrude's "human story," a vital part of the social history of the oil sands and one that has not previously been told or appreciated.

Former Syncrude President Jim Carter threw light on Syncrude's maturation. David Collyer gave me perspective on the oil industry's leadership. Adriana Davies, who was deeply involved in the Glenbow Foundation's Oil Sands Oral History Archive, was a valued guide to that important repository of personal stories. My friend Lawrence Pitt patiently helped me understand many dimensions of the complex debate over climate change. A.W. Hyndman provided perspective on Eric Newell's work on national oil sand policy. Dennis Prouse provided valuable insights on "the Mind Field" and Robert Spragins and members of the Spragins family helped with recollections of their father and family photographs of the man. Finally, I deeply appreciate the support of my wife Jennifer Shifrin, who gave me much wise counsel on the tone and substance of the book.

I also want to thank Paul Carlucci, my copy editor, for identifying many issues, and Wendy

Theodore, my formatting editor, for diligently getting my manuscript ready for Amazon publication. Any errors that remain are mine.

Engineers: An Appreciation

A book about the oil sands industry -- and all the other industries that feed us, keep us warm, puts roofs over our heads and transport us around the world -- would not be complete without an appreciation of all the engineers and technicians and tradespeople – thousands and thousands of them over the years – who designed and built what Vaclav Havel described as "the great clanking, stinking machinery "of the oil sands and made it work for the benefit of the rest of us.

Fifty years ago – I can't believe I am saying this -- I was hired by the still very-young Syncrude organization as it prepared to start into history. I was the 102nd person to be hired by Syncrude, and for a period of time I was the only person in its management who was neither an engineer nor technologist.

Those teammates weren't the easiest people to work with, and for a time they made it clear they had little regard for someone who was not "one of them." Heck, I couldn't talk about calculus! They weren't always easy to love. Their humour was politically incorrect. They were inclined to characterize anyone who disagreed with them as "an idiot." They felt that society did not appreciate them, and they had a point. The rest of the world – people like me – tended to take them, and their work, for granted, and had little appreciation for how difficult and vital it was. Not until I walked alongside them (sometimes in their moccasins) for the five years that it took to design and build and "start up" Syncrude, did I fully appreciate the value of their minds and the near-miraculous things they were capable of doing. And so I want to thank them for their contribution then and today and tomorrow.

Whether you want an energy system that works, or a trip to Mars, you won't get there without them. As an old Syncrude hand, Carl Sherman, used to say, *God love 'em.*

John J. Barr
Craig Bay Seaside Village, British Columbia

ABOUT THE AUTHOR

John J. Barr

John J. Barr's work as a Western historian builds on a career that moved from journalism and politics to public affairs. His interest in the past, present, and  possible futures of the Canadian oil sands was first fueled by his work as communication head of Syncrude Canada between 1972 and 1986. Syncrude was the proving ground for the modern oil sands industry. Barr is one of last survivors of its original management team.

At 79, he divides his time between hiking, stargazing, and historical research. He lives with his wife Jennifer Shifrin in the seaside community of Craig Bay, British Columbia, and can be reached through his website at www.johnjbarr.com.

Photo courtesy of Ernest von Rosen.